Disability in the Ottoman Arab World, 1500–1800

Physical, sensory, and mental impairments can influence an individual's status in society as much as the more familiar categories of gender, sexual orientation, age, class, religion, race, and ethnicity. This was especially true of the early modern Arab Ottoman world, where being judged able or disabled impacted every aspect of a person's life, including performance of religious rituals, marriage, job opportunities, and the ability to buy and sell property. Sara Scalenghe's book is the first on the history of both physical and mental disabilities not only in the Middle East and North Africa, but also in the premodern non-Western world. Unlike previous scholarly works that examine disability as discussed in religious texts, this study focuses on representations and classifications of disability and impairment across a wide range of primary sources, including chronicles, biographies, the law, medicine, belles lettres, and dream manuals. As such, this is a sociocultural history that seeks to explain how blindness, deafness and muteness, impairments of the mind, and intersex were understood and experienced in a specific Arab-Islamic context within the geographical area that includes present-day Egypt, Syria, Lebanon, Jordan, and Palestine/Israel under Ottoman rule in the early modern period.

Sara Scalenghe is an assistant professor of history at Loyola University Maryland.

Disability in the Ottoman Arab World, 1500–1800

SARA SCALENGHE
Loyola University Maryland

CAMBRIDGE
UNIVERSITY PRESS

CAMBRIDGE
UNIVERSITY PRESS

32 Avenue of the Americas, New York, NY 10013-2473, USA

Cambridge University Press is part of the University of Cambridge.

It furthers the University's mission by disseminating knowledge in the pursuit of education, learning, and research at the highest international levels of excellence.

www.cambridge.org
Information on this title: www.cambridge.org/9781107044791

First published 2014

Printed in the United States of America

A catalog record for this publication is available from the British Library.

Library of Congress Cataloging in Publication Data
Scalenghe, Sara, 1970– author.
Disability in the Ottoman Arab world, 1500–1800 / Sara Scalenghe, Loyola University Maryland.
pages cm. – (Cambridge studies in Islamic civilization)
ISBN 978-1-107-04479-1 (hardback)
1. People with disabilities – Middle East – History. 2. Human body – Social aspects – Middle East. 3. Intersexuality – History. 4. Insanity (Law) – Middle East – History. I. Title. II. Series: Cambridge studies in Islamic civilization.
HV1559.M53S23 2014
305.9080956–dc23 2013039679

ISBN 978-1-107-04479-1 Hardback

To my father and my brother, Franco and Davide, and to the memory of my mother and my sister, Laura and Cecilia.

Contents

Acknowledgments

This book first began as a doctoral dissertation at Georgetown University under the supervision of the incomparable Judith Tucker, the most marvelous mentor a graduate student could hope for. Throughout the years she has been unfailingly supportive, and a lot of fun to boot. I am also very grateful to the other members of my dissertation committee, Carol Benedict, Ahmad Dallal, Beshara Doumani, and John Voll, for their probing questions and incisive comments, and to Chris Toensing and Kevin Martin, who have shared this journey with me since our old days on Ontario Road in graduate school. That they read every word of the book manuscript is just one more reason I love them.

My quest for sources, especially manuscripts, in the archives and libraries of Syria, Lebanon, Egypt, Germany, the United Kingdom, Ireland, and the United States was supported by generous funding, including an International Dissertation Research Fellowship from the Social Science Research Council, a Fulbright-Hays Doctoral Dissertation Research Abroad Program, a Cosmos Club Foundation Young Scholars Award, and a Bernadotte E. Schmitt Grant for Research in European, African, or Asian History awarded by the American Historical Association.

Dissertation writing was made possible by a Charlotte W. Newcombe Dissertation Fellowship from the Woodrow Wilson National Fellowship Foundation and by a Royden B. Davis Dissertation Fellowship provided by Georgetown's History Department. A Qatar

Post-Doctoral Fellowship at the Center for Contemporary Arab Studies at Georgetown University and junior sabbatical semesters as well as summer faculty grants from Indiana University Bloomington and Loyola University Maryland were instrumental in giving me the time and resources to make substantive revisions to the manuscript, including expanding its geographical scope to Egypt.

I owe an enormous intellectual debt to Susan Burch, who introduced me to the richness of the field of disability history, critiqued each chapter of this book, and encouraged me to make it accessible to non–Middle Eastern specialists. I continue to be inspired by her integrity and the rigor of her scholarship. Jeff Brune, too, generously read every word and gave me extensive and incisive feedback that compelled me to rewrite entire sections. Cathy Kudlick's suggestions, especially for Chapter 2, were invaluable and led me to rethink some of my central assumptions. I also thank her for inviting me to share my work in her seminar on the body at the Paul K. Longmore Institute on Disability at San Francisco State University and at the Bay Area Disability Studies Consortium in March 2013. Although my flight was canceled due to stormy weather, her class discussed the book manuscript and took the trouble to type and email their insightful remarks to me.

Over the years I have benefited from the opportunity to share parts of my work in progress at conferences, workshops, seminars, graduate colloquia, and talks in the United States and abroad. In particular, I wish to thank Jim Gelvin and the Center for Near Eastern Studies at UCLA for inviting me to give a talk in February 2012, and Susan Burch for the opportunity to speak at Middlebury College in April 2013. The observations made by faculty and students at those two events were crucial for sharpening my thinking in the final stages of revising.

I am grateful to the anonymous reviewers at Cambridge University Press for their very helpful comments, to the production team at Cambridge University Press, to Nancy Zibman for preparing the index, to Laurie King for reviewing the proofs, and to my saintly father for double-checking that the citations in every footnote and in the bibliography were formatted correctly.

Note on Transliteration, Personal Names, Dates, and Translations

I have transliterated Arabic and Turkish words in accordance with the system devised by *The International Journal of Middle East Studies*. As a general rule, words that recur frequently, like *khunthā*, have diacritical marks and are italicized only the first time they appear.

As I explain in Chapter 2, Arabic personal names can be very long because they often comprise onomastic chains that report the name of the father and, if known, of the paternal grandfather and great-grandfather, in addition to several other nouns and adjectives that indicate place of origin and/or residence, occupation, honorific titles, and other distinguishing characteristics. For the sake of prose, as a general rule I provide only the first and last name of an individual in the main body of the text. Thus, for example, Muhammad b. Ahmad b. Muhammad al-Naqshabandi al-Hanafi al-Asamm al-'Arbili is shortened to Muhammad al-'Arbili. In the case of authors, a longer version of their names is listed in the footnotes and bibliography. I have, however, made several exceptions to this rule, including when a person was only known by his or her first name and father's name, for example, 'Itban b. Malik ("b." is an abbreviation for "ibn," "son of"), or for those who were commonly addressed by an honorific title rather than by their first name, for example, Najm al-Din ("the star of the faith") al-Ghazzi.

I used conversion tables to convert all dates from the Islamic calendar (called "Hijri," which means "of the emigration," because it begins in 622 CE when the Prophet Muhammad emigrated from Mecca to Medina) to the Gregorian or Common Era calendar (CE). Because the Islamic calendar is lunar and the Gregorian is solar, we can convert dates with precision only if we know the day, month, and year that an event took place. Thus, for instance, Muharram 1, 1208, Hijri corresponds to August 8, 1793, but Ramadan 1, 1208, is April 2, 1794. When the full Hijri date was not available to me and all I had was "the year 1208 Hijri," I converted it as 1793/4 CE.

All English translations from Arabic, Turkish, French, and Italian are my own unless otherwise specified.

Introduction

Disability and Its Histories in the Arab World

This book explores what it meant to have disabilities in the Ottoman Arab world. The first stumbling block in the path of anyone wishing to study disability in the Arab world is the absence of an Arabic word for "disability" until fairly recently. "The disabled" and "disability" as discrete categories are products of modern Europe.[1] They did not appear in the Arab world until the twentieth century, and it is thus not surprising that the current Arabic equivalents of the English words "disability" (*iʿāqa)* and "disabled" (*muʿawwaq*; also *maʿūq* and *muʿāq*) are of correspondingly recent coinage. They both derive from the verb *ʿāqa*, "to hinder" or "to hamper."[2]

This does not mean, of course, that awareness of impairments did not exist prior to the modern period. In fact, a distinctive characteristic of Arabic literature well into the eighteenth century is works or chapters of works devoted to "people with defects" or "people with blights," variously referred to as *ahl al-ʿāhāt*, *aṣḥāb al-ʿāhāt*, or *dhawū al-ʿāhāt*. Kristina Richardson has noted that the word *ʿāhā*, "blight," "defect," or "damage," "is not a word intrinsically linked to the human body – it could also be applied to crops, for instance – and does not

[1] On the origin of the category of "disability," see Lennard Davis's classic study, *Enforcing Normalcy: Disability, Deafness, and the Body* (London; New York: Verso, 1995).

[2] Saʿdi Abu Habib, *al-Muʿawwaq wa-l-Mujtamaʿ fi al-Shariʿa al-Islamiyya* (Damascus: Dar al-Fikr, 1982), 12.

necessarily connote physical disability. Rather, it suggests a mark that spoils the presumed wholeness, integrity, or aesthetics of something."[3] Hence, "blights" encompassed a wide range of mostly physical traits as varied as blindness, leprosy, flat noses, and even blue eyes.

One of the earliest and best-known representatives of this genre is *The Book of the Leper, Lame, Blind, and Cross-Eyed* by the famous Abbasid belles-lettrist al-Jahiz (d. 868/9). It addresses a wide array of physical defects, from skin disorders to lameness, goggle-eyes, hemiplegia (paralysis of one side of the body), leprosy, hunchback, scrotal hernia, warts, absence of eyebrows, short necks, and baldness.[4]

In the centuries that followed, other Arab authors followed this literary trend.[5] To mention but two examples from the Ottoman era, the Meccan Hadith scholar Muhammad al-Makki (d. 1547) wrote a treatise on people with physical "blights" that described men who were blind, cross-eyed, had hunched backs, or were bald. Al-Makki gathered information from the Hadith, biographies, and various anecdotes, and stated that his purpose was edification and entertainment, not slander.[6] The entertaining aspect of this type of literature is even more salient in the work of the Egyptian belles-lettrist Ahmad al-Hifnawi al-Bishari (fl. 1769), which contains anecdotes, poems, jokes, and stories about judges, poets, doctors, the insane, thieves, and many others. One of al-Bishari's chapters is devoted to "people with blights": those who were blind, one-eyed, cross-eyed, had hunched backs, were bald, those suffering from bad breath, and those with very long beards.[7]

3 Kristina L. Richardson, "Disability? Perspectives on Bodily Difference from the Middle East," in *Why the Middle Ages Matter: Medieval Light on Modern Injustice*, ed. Celia Ghazelle et al., 121–129 (London; New York: Routledge, 2012), 125. Richardson discusses the etymology of *'āhā* and its uses in medieval Arabic sources more extensively in *Difference and Disability in the Medieval Islamic World: Blighted Bodies* (Edinburgh: Edinburgh University Press, 2012).

4 Abu 'Uthman al-Jahiz, *Kitab al-Bursan wa-l-'Umyan wa-l-Hulan* [The Book of the Leper, Lame, Blind, and Cross-Eyed], ed. Muhammad Mursi al-Khuli (Beirut: Mu'assasat al-Risala, 1981).

5 Richardson, *Difference and Disability*, especially 9–13.

6 Jar Allah Muhammad b. Fahd al-Makki, *al-Nukat al-Ziraf fi al-Maw'iza bi-Dhawi al-'Ahat min al-Ashraf*, MS, Chester Beatty Library, Dublin, Ar. 3838.

7 Shihab al-Din Ahmad al-Hifnawi al-Bishari, *Bughyat al-Jalis wa-l-Musamir wa-Nuzhat al-Arwah wa-l-Khawatir*, MS, Chester Beatty Library, Dublin, Ar. 4853. I was unable to find biographical information on al-Bishari beyond the fact that he was

It is evident, then, that the category of "people with blights" is not synonymous with "people with disabilities." It encompasses a much broader range of conditions than today's "impairments" and "disabilities," which are intrinsically connected to individual performance and productivity. Some, like hemiplegia, fall within our modern conception of disability; others, such as flat noses, bad breath, or freckled faces, do not. In addition, because it refers to visible defects that mark the body, *'āhā* excludes physical, cognitive, sexual, and reproductive impairments that are difficult or impossible to identify by sight.

To complicate matters further, the phrase "people with blights" only rarely appears in nonliterary sources like medicine and the law. One possible explanation is that many "blights," such as short necks and very long beards, did not require or were not amenable to medical treatment and did not affect a person's legal status in any way. Islamic law, for its part, employed the words *zamāna*, "chronic illness" or "chronic condition," and *zamin* or *zamīn* for "someone with a chronic illness or condition" in certain contexts in reference to physical impairments like missing limbs, partial paralysis, muteness, and blindness. One context was work: a chronic condition might prevent a man from earning a living, which would entitle him to financial support from his relatives (this was a moot point for women, because according to Islamic law, they ought to be supported financially by their fathers, husbands, or other male relatives).

Although there was no category for "the disabled," either conceptually or linguistically, people with impairments of the body and of the mind are everywhere in Arabic sources dating from Ottoman times. The reason for this striking frequency may be, at least in part, a function of the sheer numbers of people with impairments who lived in Arab towns and cities. People acquired impairments for much the same reasons that people acquired them all over the world: war, poverty, congenital anomalies, injuries, illnesses, and diseases, including the waves of bubonic plague that ravaged Greater Syria and Egypt.

But there are at least two additional causes that are particular to the region. The first is consanguineous marriages, also known as "cousin

alive as late as 1773, when the autograph copy of this manuscript was completed (the original was composed in 1769).

marriages," which are associated with a relatively high incidence of certain congenital malformations and of conditions caused by autosomal recessive genes, including hereditary deafness, certain types of blindness and muscular dystrophies, neurological disorders, and disorders of sex development. Historically, marriage between cousins has been practiced widely throughout the Arab world and Southwestern Asia. Cousin marriage occurs most commonly among paternal first cousins, including double first-cousin marriage (two siblings who marry two siblings who are their first-degree cousins). It is a cultural, not religious, norm that may be rooted historically in the desire to keep property within the family. While more common among Muslims, it is practiced by non–Muslims as well. The rate of consanguinity in the Arab world is one of the highest in the world: clinical geneticists estimate that to this day an average of 40 percent of marriages in many Arab countries are consanguineous, and that the percentage may be as high as 90 percent in certain Bedouin communities.[8]

Second, the emerging field of medical geology is beginning to reveal surprising clues about the relationship between geological features and the prevalence of certain impairments in the Middle East. Iodine deficiency is a case in point. Iodine is a chemical element that is needed for the correct functioning of the thyroid glands. It is not produced by the human body and can only be acquired through food. An insufficient supply of iodine can cause goiter (enlargement of the thyroid) and, in the children of women who had iodine deficiency during pregnancy, deafness, stunted growth, and damage to brain development. In fact, iodine deficiency is the most common cause of preventable congenital cognitive disabilities in the world. Medical geologists have discovered that the main origin of the deficiency in parts of Lebanon and Syria is that the soil itself does not retain iodine, which is thus not bioavailable.[9] It follows that iodine deficiency and the impairments it produces were historically endemic in the Eastern Mediterranean region well into the second half of the twentieth century, when

[8] Ahmad S. Teebi, ed., *Genetic Disorders among Arab Populations*, 2nd ed. (Berlin: Springer, 2010), 6. See also Ghazi O. Tadmouri et al., "Consanguinity and Reproductive Health among Arabs," *Reproductive Health* 6, no. 17 (2009): n.p.

[9] Humam Misconi and Maryam Navi, "Medical Geology in the Middle East," in *Medical Geology: A Regional Synthesis*, ed. Olle Selinus, Robert B. Finkelman, and Jose A. Centeno (Dordrecht, The Netherlands; London: Springer, 2010), 149–156.

the deficiency was recognized as a serious public health issue and salt started to be iodized.

Although people with impairments were an integral part of the social landscape of the Ottoman Arab world, they have been almost entirely ignored by scholars. Even social historians, who are usually sensitive to the lives of the marginalized and the disempowered, and who have written about most facets of the human experience in depth and with nuance, have hardly displayed any interest in disability. Indeed, the pithy observation by U.S. disability historian Douglas Baynton that "[d]isability is everywhere in history, once you begin looking for it, but conspicuously absent in the histories we write," applies perfectly to the scholarship on the Arab world.[10] To the best of my knowledge, the only book-length historical study of disability anywhere in the Arab lands in any period and in any language is Kristina Richardson's recently published *Difference and Disability in the Medieval Islamic World: Blighted Bodies* (2012). It is largely a literary history centered on the writings of six male Sunni Muslim scholars in fifteenth- and sixteenth-century Cairo, Damascus, and Mecca who wrote about "blighted bodies."[11]

While not about the Arab world, M. Miles's study of the deaf at the Ottoman court (2000) and Ayşe Ezgi Dikici's MA thesis, "Imperfect Bodies, Perfect Companions? Dwarfs and Mutes at the Ottoman Court in the Sixteenth and Seventeenth Centuries" (2006), are useful studies of disability in Ottoman Turkey. Firoozeh Kashani-Sabet's pioneering article on the history of disability in nineteenth- and twentieth-century Iran (2010) offers interesting insights for comparative purposes, and the late Michael Dols's *Majnūn: The Madman in Medieval Islamic Society* (1992), remains a classic despite the fact that it was not written from the perspective of disability studies.[12]

[10] Douglas C. Baynton, "Disability and the Justification of Inequality in American History," in *The New Disability History: American Perspectives*, ed. Paul K. Longmore and Lauri Umansky (New York: New York University Press, 2001), 51.

[11] Richardson, *Difference and Disability*. See also her article titled "Disability?" Despite its title, Fareed Haj's *Disability in Antiquity* (New York: Philosophical Society, 1970), where antiquity is defined as the period spanning from the rise of Islam to 1258 CE, is primarily a medical history and is dated in terms of sources, approach, and methodology.

[12] M. Miles, "Signing in the Seraglio: Mutes, Dwarfs and Jestures at the Ottoman Court 1500–1700," *Disability and Society* 15, no. 1 (2000): 115–134; Ayşe Ezgi Dikici,

Although not works of history, of much value to historians are Fedwa Malti-Douglas's groundbreaking article on blindness in the medieval period (1989), Yusuf Sadan's study of "blights" in Arabic literature (1983), Mohammed Ghaly's a meticulously researched book on disability in the writings of prominent Muslim theologians and jurists from early Islamic times to the present day Vardit Rispler-Chaim's legal history entitled *Disability in Islamic Law* (2007), and Muhammad Hawwa's study of "people with special needs" in Islamic law (2010).[13]

The literature on urogenital anomalies and hermaphroditism, better known today as intersex or disorders of sex development, is even more sparse and is focused almost exclusively on Islamic legal theory: Agostino Cilardo's "Historical Development of the Legal Doctrine Relative to the Position of the Hermaphrodite in the Islamic Law" (1986); Ulrich Rebstock's "Mathematischen Quellen zur Rechtsgeschichte: Das Problem des Hermaphroditen" (1990), which charts the complex mathematical formulas used by jurists to determine a hermaphrodite's share of inheritance; Paula Sanders's "Gendering the Ungendered Body: Hermaphrodites in Medieval Islamic Law" (1991);

"Imperfect Bodies, Perfect Companions? Dwarfs and Mutes at the Ottoman Court in the Sixteenth and Seventeenth Centuries," Master's thesis, Sabancı University, 2006; Firoozeh Kashani-Sabet, "The Haves and the Have Nots: A Historical Study of Disability in Modern Iran," *Iranian Studies* 43, no. 2 (2010): 167–195; Michael Dols, *Majnūn: The Madman in Medieval Islamic Society*, ed. Diana E. Immisch (Oxford: Clarendon Press, 1992). See also M. Miles's extensive bibliography titled *Disability and Deafness in the Middle East: A Bibliography Comprising Materials with Technical, Cultural and Historical Relevance to Child and Adult Impairments, Disabilities and Deafness, Incapacity, Mental Disorders, Special Needs, Social and Educational Responses and Rehabilitation; Partly Annotated*. 2008. Available online at http://cirrie.buffalo.edu/bibliography/mideast/index.php. Hosted by the Center for International Rehabilitation Research Information and Exchange. Accessed on July 13, 2013.

13 Fedwa Malti-Douglas, "*Mentalités* and Marginality: Blindness and Mamlûk Civilization," in *The Islamic World from Classical to Modern Times: Essays in Honor of Bernard Lewis*, ed. Clifford Edmund Bosworth et al. (Princeton, NJ: Darwin Press, 1989), 211–238; Yusuf Sadan, *al-Adab al-ʿArabi al-Hazil wa-Nawadir al-Thuqala': al-ʿAhat wa-l-Masawi' al-Insaniyya wa-Makanatuha fi al-Adab al-Raqi* (Köln, Germany: Al-Kamel Verlag, 2007. First published in 1983); Vardit Rispler-Chaim, *Disability in Islamic Law* (Dordrecht, The Netherlands: Springer, 2007); Mohammed Ghaly, *Islam and Disability: Perspectives in Theology and Jurisprudence* (London; New York: Routledge, 2010); Muhammad b. Mahmud Hawwa, *Huquq Dhawi al-Ihtiyajat al-Khassa fi al-Shariʿa al-Islamiyya* (Beirut: Dar Ibn Hazm li-l-Tibaʿa wa-l-Nashr wa-l-Tawziʿ, 2010).

and, more recently, a brief chapter in Vardit Rispler-Chaim's *Disability in Islamic Law* (2007). Rispler-Chaim is the only scholar who has approached intersex as a disability.[14]

This dearth of scholarship is not unique to the Arab world or the Middle East but extends to all other non–Western societies, both premodern and modern. To the best of my knowledge, with the exception of Kristina Richardson's *Difference and Disability in the Medieval Islamic World* (2012), there are no histories of disability before the nineteenth century in any part of the world besides Europe and the United States. Disability history in the West, on the other hand, has made spectacular strides and is now a vibrant field, tackling topics as diverse as blindness in nineteenth-century France, "ugly laws" in the United States (laws that as late as 1974 ordered people with disabilities regarded as unsightly not to appear in public), eugenic programs in Nazi Germany, war veterans, prosthetics, freak shows, and much else in between.[15] In the last few years a cluster of publications has centered on disability in the European Middle Ages, a welcome and promising antidote to the general presentism of the field.[16]

[14] These articles and chapters were published, respectively, in *The Search: Journal for Arab and Islamic Studies* 7 (1986): 128–170; *Die Welt des Orients* 20–21 (1990): 99–114; in *Women in Middle Eastern History: Shifting Boundaries in Sex and Gender*, ed. Nikki R. Keddie and Beth Baron (New Haven, CT: Yale University Press, 1991), 74–95; *Disability in Islamic Law*, 69–75. Abdelwahab Bouhdiba devotes about three pages of his *Sexuality in Islam* to hermaphrodites. He assumes that Muslim jurists were greatly embarrassed when faced with cases of hermaphroditism. His discussion, in addition to being cursory, is flawed by a misunderstanding of the Arabic terminology pertaining to hermaphrodites, whereby he conflates the term and concept of *khunthā* (hermaphrodite) with that of *mukhannath* (effeminate). Abdelwahab Bouhdiba, *Sexuality in Islam*, translated from French by Alan Sheridan (London: Saqi Books, 1998. Originally published in 1975), 40–42.

[15] Although more than ten years old, Catherine Kudlick's review article "Disability History: Why We Need Another 'Other,'" *The American Historical Review* 108, no. 3 (2003): 763–793 still offers one of the clearest and most comprehensive overviews of the development of the field.

[16] Irina Metzler is one of the most prolific scholars in this field. See *A Social History of Disability in the Middle Ages: Cultural Considerations of Physical Impairment* (New York: Routledge, 2013); *Disability in Medieval Europe: Thinking About Physical Impairment During the High Middle Ages, c. 1100–1400* (London: Routledge, 2006); "Disability in the Middle Ages: Impairment at the Intersection of Historical Inquiry and Disability Studies," *History Compass* 9, no. 1 (2011): 45–60. See also Joshua R. Eyler, ed., *Disability in the Middle Ages: Reconsiderations and Reverberations* (Burlington, VT: Ashgate, 2010); Wendy Turner and Tory Vandeventer Pearman, eds., *The Treatment of Disabled Persons in Medieval Europe: Examining Disability*

The need for histories of disability in non–Western contexts is particularly urgent if we aspire to avoid Euro-American centrism, a narrative that is circumscribed by the experiences of Western Europe and the United States. To quote Helen Meekosha, "contemporary disability studies constitutes a form of scholarship colonialism"[17] whose implications extend well beyond academia. An estimated 80 percent of the world's disabled people live in the Global South. Although usually well meaning, the ways that agencies such as the United Nations, the World Bank, the International Monetary Fund, USAID, and nongovernmental organizations such as Oxfam choose to frame and finance disability-related projects are almost invariably predicated on Western models of rehabilitation and development based on liberal philosophies, with little knowledge of different non–Western local community practices.

This "disability imperialism," as we might call it, is bitterly ironic when millions of people in the Global South become disabled as a result of the appalling conditions of many of the sweatshops that produce cheap manufactured goods for the North, or as the direct or indirect consequence of the wars waged by the United States and its allies. For example, recent medical studies have exposed some unexpected repercussions of the U.S.-led wars in Iraq. It is widely suspected that depleted uranium was used in the 2004 bombings of Fallujah. The rate of cancers, congenital anomalies in newborns, including ambiguous genitalia, and other health problems since then has soared dramatically, possibly exceeding that of the survivors of the U.S. nuclear bombing of the Japanese cities of Hiroshima and Nagasaki in 1945.[18]

in the Historical, Legal, Literary, Medical, and Religious Discourses of the Middle Ages (Lewiston, NY: The Edwin Mellen Press, 2010); Wendy Turner, ed., *Madness in Medieval Law and Custom* (Leiden: Brill: 2010); David Turner, *Disability in Eighteenth-Century England: Imagining Physical Impairment* (New York; London: Routledge, 2012); Edward Wheatley, *Stumbling Blocks Before the Blind: Medieval Constructions of a Disability* (Ann Arbor: The University of Michigan Press, 2010).

17 Helen Meekosha, "Decolonising Disability: Thinking and Acting Globally," *Disability and Society* 26, no. 6 (October 2011): 668. See also Helen Meekosha and Karen Soldatic, "Human Rights and the Global South: The Case of Disability," *Third World Quarterly* 32, no. 8 (2011): 1383–1398.

18 Samira Alaani et al., "Pilot Study of Congenital Anomaly Rates at Birth in Fallujah, Iraq, 2010," *Journal of the Islamic Association of North America* 44 (2012): 1–7; Chris Busby, Malak Hamdan, and Entesar Ariabi, "Cancer, Infant Mortality and Birth-Sex Ratio in Fallujah, Iraq, 2005–2009," *International Journal of*

In this context, the strategies employed by disability rights activists in the United States are not necessarily the most appropriate outside North America. Take, for example, Disability Pride, a movement grounded in minority identity politics and inspired by Black Pride, Gay Pride, and others. Whereas Disability Pride has been effective and empowering in the United States, it may be a less beneficial tool in locales where focusing on prevention is more urgently needed than focusing on promoting pride.

Scholarship on the contemporary Arab world highlights the dire conditions in which many people with disabilities in the region live. Negative attitudes toward disability are widespread, and, as in many other parts of the world, women and people with intellectual disabilities suffer double discrimination.[19] The problem with some of this scholarship is that it displays the tendency to project backward in time today's relatively low social and economic status of disabled people and to assume that it must be the relic of the past. Some writers impute this lamentable state of affairs to the supposedly discriminatory statements of the foundational texts of Islam, the Qur'an and the Hadith.[20]

Environmental Research and Public Health 7 (2010): 2828–2837. See also Patrick Cockburn, "Toxic Legacy of U.S. Assault in Fallujah 'Worse Than Hiroshima'," *The Independent*, July 24, 2010. Available online at http://www.independent.co.uk/news/world/middle-east/toxic-legacy-of-us-assault-on-fallujah-worse-than-hiroshima-2034065.html. Accessed July 9, 2013.

[19] In 2007 *The Review of Disability Studies: An International Journal* dedicated an entire issue to the Middle East (vol. 3, no. 4). See also Amira Abd El-Khalek, "The Cultural Construction of Women with Disabilities in Egypt: An Ethnographic Approach," *The MIT Electronic Journal of Middle East Studies* 4 (Fall 2004): 92–107; Lesley Lababidi, in collaboration with Nadia El-Arabi, *Silent No More: Special Needs People in Egypt* (Cairo; New York: The American University in Cairo Press, 2002). Lina Abu-Habib reports a particularly chilling case. In 1993, during an Israeli military offensive in South Lebanon, a father abandoned his physically disabled teenage daughter, leaving her helpless amid the combat occurring around their home. He had not failed to rescue the family cow, however, reasoning that it would be more useful to the family. After a neighbor reported the incident to a local NGO, the girl was rescued and the father was asked to take her to the hospital. He refused, protesting "What for? So that I start paying for her?" In Lina Abu-Habib, ed., *Gender and Disability: Women's Experiences in the Middle East* (Oxford: Oxfam, 1997), 4.

[20] For an example of this attitude, see Majid Turmusani's *Disabled People and Economic Needs in the Developing World: A Political Perspective from Jordan* (Aldershot, UK: Ashgate, 2003) and his "Disabled Women in Islam: Middle Eastern Perspective," *Journal of Religion, Disability and Health* 5, no. 2/3 (2001): 73–85.

Given the historiographical near void on disability in the Middle East, however, these backward projections are simply not warranted. This book hopes to contribute a historically grounded perspective to this conversation.

Framing This Book

To the extent that the sources permit it, I am interested in recovering and documenting the lived experiences of people who had impairments. How did Arabs view and treat people with impairments? In what ways did impairments disable individuals? Were impairments fundamental constituents of a person's identity? How did they inform social, economic, and reproductive relations? Were people with impairments ostracized, marginalized, or integrated? Were their conditions attributed to divine punishment or to moral failing? What role did religion play in the construction of and attitudes to impairments? In what ways were the experiences of impairment influenced by gender and by socioeconomic class?

Like most historians of disability today, I believe that disability is as essential a category of analysis as the more familiar triad of race, class, and gender. And like many historians of disability today, I take a social constructionist approach that employs the "social model" of disability. The "social model" replaces the earlier "medical model," according to which disabilities are medical problems that reside within the individual and, as such, are pathologies in need of correction and rehabilitation, existing independently of historical processes and cultural practices. Earlier histories of specific diseases (leprosy, for example) were usually framed within the medical model of disability.

Some scholars maintain that the social model is premised on a distinction between impairment and disability. An impairment is whatever a given society regards as an anatomical abnormality or physiological loss of function and is built on a belief about function, whereas disability is the systemic societal response to perceived impairments. In other words, while impairments are not transhistorical, they are more "objective" than disability within a given society. Thus in the United States in the twenty-first century, nearsightedness (myopia), a physiological condition of the eye, can be considered an impairment but only becomes a disability if corrective lenses or corrective surgery are not available.

A powerfully clear illustration of how disability is socially constructed is still the parable written in 1975 by Victor Finkelstein, an anti-apartheid activist and one of the early leaders of the disability rights movement first in South Africa and then in the United Kingdom. Finkelstein asks us to imagine a village inhabited almost entirely by wheelchair users who run it according to their needs. What happens when able-bodied people move into the village is worth quoting at length:

> They design their own buildings to suit their physical situation. One thing the wheelchair-user architects quickly discover in this village is that because everyone is always in wheelchairs there is no need to have ceilings at 9 feet 6 inches high or door heights at 7 feet 2 inches. Soon it becomes standard practice to build doors to a height of 5 feet and ceiling or rooms to a height of 7 feet 4 inches. Naturally the building codes set out in the regulations made these heights standard. Now everyone is happy in the village; all the physical difficulties have been overcome and this little society has changed according to the physical character of its members. At last the buildings and environment are truly in tune with their needs.
>
> Let us say that when all the adjustments had been made and became fixed, in this wheelchair-user society, a few able-bodied had, through no choice of their own, to come and settle in this village. Naturally, one of the first things they noticed was the heights of the doors and ceilings. They noticed this directly, by constantly knocking their heads on the door lintels. Soon all the able-bodied members of the village were also marked by the dark bruises they carried on their foreheads. Of course, they went to see the village doctors, who were, naturally, also wheelchair-users. Soon the wheelchair-user doctors, wheelchair-user psychiatrists, wheelchair-user social workers, etc., were involved in the problems of the able-bodied villagers. The doctors produced learned reports about the aches and pains of the able-bodied in society. They saw how the bruises and painful backs (from walking bent double so frequently) were caused by their physical condition. The wheelchair-user doctors analysed the problems and wrote their definitions. They said these able-bodied people suffered a "loss or reduction of functional ability" which resulted in a handicap. This handicap caused a "disadvantage or restriction of activity" which made them disabled in this society.
>
> Soon special aids were designed by the wheelchair-user doctors and associated professions for the able-bodied disabled members of the village. All the able-bodied were given special toughened helmets (provided free by the

village) to wear at all times. Special braces were designed which gave support while keeping the able-bodied wearer bent at a height similar to their fellow wheelchair-user villagers. Some doctors went so far as to suggest that there was no hope for these poor sufferers unless they too used wheelchairs, and one person even went so far as to suggest amputation to bring the able-bodied down to the right height. The able-bodied disabled caused many problems. When they sought jobs no one would employ them. Special experts had to be trained to understand these problems and new professions created for their care....

In time special provision had to be made in the village to provide a means of obtaining money for these able-bodied disabled to live. Voluntary societies were created to collect charity and many shops and pubs had an upturned helmet placed on the counters for customers to leave their small change. Painted on the helmets were the words "Help the able-bodied disabled." Sometimes a little plaster-cast model would stand in the corner of a shop – the figure bent double, in their characteristic pose, with a slotted box on the figure's back for small coins.[21]

The allegory concludes with the able-bodied disabled people organizing to fight against their marginalization. Their demands for societal changes, such as raising the heights of buildings' doors and ceilings, are met with opposition by some irritated wheelchair users who believe they should stop causing problems for the village and just accept the disadvantages caused by their impairment. The moral of the tale is that a person is not inherently disabled, but rather is disabled by society.

Given that what it means to be disabled is produced by society and changes through time and place, this book does not assume that people with impairments in the Ottoman Arab world were automatically disabled. Faced with uncharted territory and a daunting plethora of possibilities, but also restricted by the availability of primary sources, I opted to limit the scope of this study to permanent impairments (or potentially permanent, like certain forms of madness) that could hinder substantially a person's ability to participate fully in some of the major spheres of life, namely the ability to make a living, take part in religious rituals, sexuality, marriage, and reproduction. The four core chapters of this book explore deafness and muteness,

[21] Victor Finkelstein, "To Deny or Not to Deny Disability," in *Handicap in a Social World: A Reader*, ed. Ann Brechin, Penny Liddiard, and John Swain (Sevenoaks, UK: Hodder and Stoughton, 1981), 34–36. Originally published in *The Magic Carpet* 27, no. 1 (1975): 31–38.

blindness, impairments of the mind, and intersex and urogenital anomalies, respectively. To some extent these are of course subjective choices, and they are inevitably shaped by my Western twenty-first-century notions of disability. But, significantly, those are also among the impairments that recur most frequently in early modern Arabic sources. Baldness and bad breath, found in Arabic belles-lettres under the heading of "people with blights," motility impairments, hemiplegia (partial paralysis), and, above all, leprosy – a chronic illness that could bear devastating legal, economic, and social consequences – are addressed only in passing because they are discussed too infrequently in the sources I examined.

Deafness, blindness, and impairments of the mind are commonly found in histories of disability. Some readers, however, may be surprised to find urogenital anomalies and intersex discussed in this book (I adopt legal scholar Julie Greenberg's definition of an intersex person as "anyone with a congenital condition whose sex chromosomes, gonads, or internal or external sexual anatomy do not fit clearly into the binary female/male norm."[22]) In fact, I am not aware of works of history that examine intersex conditions alongside physical, sensory, or mental impairments.[23] Yet in the Ottoman Arab East

[22] Julie A. Greenberg, *Intersexuality and the Law: Why Sex Matters* (New York: New York University Press, 2012), 1. Until less than 100 years ago intersex conditions were known as hermaphroditism. In the United States, starting in the 1950s the word "hermaphrodite" came under sustained attack by activists on the grounds that it was not only stigmatizing but also descriptively inaccurate, for it suggests that there are people who have both female and male sets of genitals, an impossibility in humans. The term "intersex" was adopted in its place because it was found to be more precise. Within a few decades, however, the term "intersex," too, was attacked for being only marginally less stigmatizing than "hermaphrodite," and in 2006 a new, clinically based coinage, disorders of sex development (DSD), emerged. But this new nomenclature has by no means been accepted by all. The main objection is that "disorder" implies that the condition requires medical attention and correction, which is not always the case. Elizabeth Reis proposes instead "a new term, still DSD, but standing for divergence of sex development. By using this nomenclature, doctors, patients, and parents can articulate difference but not disorder." Elizabeth Reis, *Bodies in Doubt: An American History of Intersex* (Baltimore: The Johns Hopkins University Press, 2009), xv. As of the time of writing, there is little consensus beyond the undesirability of "hermaphrodite," and both "intersex" and "disorders of sex development" are used.

[23] It was the work of Michel Foucault, specifically his edition of the memoirs of the nineteenth-century French hermaphrodite Herculine Barbin, that spurred serious historical investigations of hermaphrodites (Michel Foucault, *Herculine Barbin, dite*

ambiguous genitalia were frankly discussed, especially in legal sources, because of the problems they could engender in the sphere of marriage. Hence, a woman with a severe form of labial adhesions might have been more impaired than a man with a missing arm or leg. Complex laws were elaborated to address intersex conditions, which, as will be shown in Chapter 4, could be highly disabling not only in matters of marriage, sex, and reproduction, but in other spheres of life as well.

The geographical scopc of this book is the region of the Middle East that corresponds to today's Egypt, Palestine/Israel, Jordan, Lebanon, and Syria. I refer to this area interchangeably as the Arab East, the Ottoman Arab lands, the Arab-Islamic world, and Ottoman Syria and Egypt. Ottoman Syria, early modern Syria, Greater Syria, and the Levant all refer to the geographical area bounded by the Taurus Mountains, the Mediterranean Sea, and the Euphrates River, corresponding to the modern states of Syria, Lebanon, Israel/Palestine, Jordan, and the Turkish province of Hatay.[24] I chose Egypt and Syria because they were the core Arab provinces of the Ottoman Empire and included major foci of cultural production such as the large urban centers of Cairo, Damascus, Aleppo, and Jerusalem, all of which produced an extraordinary wealth of primary sources.

The temporal parameters span from roughly the early sixteenth century, when the Arab lands were incorporated into the Ottoman Empire, to the early nineteenth century – a stretch of time conventionally referred to as the early modern period.[25] I expected to find more differences across space and found instead more similarities, as I

Alexina B. [Paris: Gallimard, 1978]). Since then historians have focused on what the perception of and treatment accorded to hermaphrodites can tell us about notions of gender, sexuality, and identity, but not disability. Anthropologists have examined intersex conditions in non-Western societies, notably in the Dominican Republic and Papua New Guinea, but most of the historical scholarship has centered on Europe, with a particular emphasis on the early modern and modern periods. Again, intersex in the non-Western world remains almost entirely unexplored by historians.

24 This area, known to the Arabs as al-Iskandarun, was part of geographic Syria until the 1930s, when the French Mandatory authorities arranged its transfer to the Republic of Turkey.

25 The validity and utility of "early modern" in reference to the history of the Middle East is beginning to be subjected to thorough scholarly engagement, at last, but that is a debate that lies outside the scope of this book.

expected to find more differences across time and found instead continuities – a reminder that the pace of cultural change can be slow. The Epilogue addresses, albeit tentatively, some of the developments in ideas about impairments that occurred in the nineteenth and twentieth centuries.

It bears emphasizing that this is not a study of disability in "Islam." My focus is not on the foundational texts of the religion, the Qur'an and the Hadith (the sayings of the Prophet Muhammad). Although, as we shall see, these texts played a critical role in shaping attitudes toward disability, I strove to examine a diverse corpus of sources written in Ottoman Syria and Egypt in the hope of recovering the widest possible range of information, narratives, and discourses about impairments. These sources include biographical dictionaries, chronicles, travelogues, legal and medical texts, treatises on physiognomy, dream manuals, essays on specific subjects (on blindness, for example), works of belles-lettres, such as collections of jokes and anecdotes, and Arabic lexicons.

Chronicles, some of which are written in semi-autobiographical or autobiographical style, travelogues, and biographical dictionaries proved to be a treasure trove, as they offer countless descriptions of and comments about people with physical and mental impairments. Indeed, historians of the Middle East are singularly fortunate to be able to mine biographical dictionaries, an extraordinarily rich body of literature that is a "unique product of Arab Muslim culture."[26] The earliest extant example of the biographical dictionary dates from ninth-century Iraq, and by the sixteenth century it was a firmly established and well-respected form of historical writing. For reasons that are unclear, the genre appears to have reached its peak in Greater Syria under the Ottomans (hence the preponderance, in this book, of biographies of

[26] Tarif Khalidi, "Islamic Biographical Dictionaries: A Preliminary Assessment," *The Muslim World* 63, no. 1 (1973): 53. For a full examination of biographical dictionaries, see also H. A. R. Gibb, "Islamic Biographical Literature," in *Historians of the Middle East*, ed. Bernard Lewis and P. M. Holt (London: Oxford University Press, 1962), 54–58; Paul Auchterlonie, *Arabic Biographical Dictionaries: A Summary Guide and Bibliography* (Durham, UK: Middle East Libraries Committee, 1987); and Dwight F. Reynolds, ed., *Interpreting the Self: Autobiography in the Arabic Literary Tradition* (Berkeley: University of California Press, 2001), especially 36–51.

men who lived in that region), but early modern Egyptian authors produced several biographical dictionaries as well.

In thousands of biographies of varying lengths, biographers recorded the lives of men from all walks of life, often their contemporaries or one or two generations removed, whom they deemed notable in the broadest sense. The Damascus-based Hasan al-Burini (d. 1615) stands out because he included biographical sketches of some of his close male friends whose only possible claim to noteworthiness was friendship with the author. We read about rulers and other state officials, jurists, Sufis, teachers, hospital and school administrators, preachers, muezzins, mosque prayer leaders, soldiers, merchants, craftsmen, physicians, entertainers, political rebels, imposters posing as prophets, and even victims of rape or murder.

Biographers took their works very seriously. Although many entries are hagiographical, authors consistently expressed their desire to present the most accurate and fairest depiction of their subjects. To that end, they supplemented the information they gathered firsthand with a plethora of both written and oral sources they reckoned credible. A major drawback of the genre for today's historians, however, is that individual entries were rarely dedicated to women; when women do appear in Ottoman-era Arabic biographical dictionaries, it is usually as the wives, mothers, daughters, or other relatives of a male biographee.[27]

This book also draws extensively on Islamic jurisprudence. Islamic law (Shari'a) differs in significant respects from other legal systems. In the words of Joseph Schacht, a foremost legal historian, the Shari'a is "an all-embracing body of religious duties," in the sense that "it comprises on an equal footing ordinances regarding worship and ritual, as well as political and (in the narrow sense) legal rules," so that "it is impossible to understand Islam without understanding Islamic law."[28] Most Muslims in the Arab lands of the Ottoman Empire belonged to four Sunni schools of law: the Hanafis, Shafi'is, Malikis, and Hanbalis.

[27] This was not the case in pre-Ottoman biographical dictionaries, which included entries on women. We do not know what caused this change in Ottoman times. See Ruth Roded, *Women in Islamic Biographical Collections: From Ibn Sa'd to Who's Who* (Boulder, CO: Lynne Rienner Publishers, 1994).

[28] Joseph Schacht, *An Introduction to Islamic Law* (Oxford: Clarendon Press, 1964), 1.

I concentrate on the Hanafi school because it was the official school of law of the empire, but I also pay attention to the Shafi'is, who had many adherents in Egypt and Syria, and to a lesser extent to the Maliki and Hanbali schools, mostly prevalent in North Africa and in the Arabian Peninsula, respectively, but with pockets of followers in the Arab East.[29]

I use two main types of legal sources: law books and collections of fatwas. Law books lay down rules and regulations for almost every aspect of every Muslim's life from the cradle to the grave, while fatwas are nonbinding but authoritative legal opinions issued by a mufti, that is, a person qualified to issue fatwas, in response to specific questions. Because questions were posed by members of all socioeconomic strata of society (the service was supposed to be free of charge), fatwas not only afford us a glimpse into the process by which scholars interpreted the law "on the ground," so to speak, but also represent a barometer for the legal and social issues that were of concern at any given time in any given community of Muslims. As we shall see in the chapters that follow, impairments featured prominently in Islamic law.

This book is not a history of medicine, but I provide some medical background because ideas about impairments were shaped in no small part by an understanding of the body rooted in humoral medicine. The Ottoman period was a time of medical pluralism characterized by the coexistence and overlapping of three different medical traditions: humoral medicine, religious medicine, and folk medicine.[30] Humoral medicine, the dominant medical framework both in the Middle East and in Europe until the nineteenth century, was grounded in the Greek theory of the four humors. According to this theory, the human body consisted of blood, black bile, yellow bile, and phlegm. A person's health depended on the correct balance of these four substances. Religious medicine or "Prophetic medicine" relied on the sayings and actions of the Prophet Muhammad concerning health,

[29] Upper Egypt was predominantly Maliki. The Hanafi school was the official school of law of the Ottomans, but Muslim subjects were free to follow whichever school they wanted, and at least the major cities in the Arab lands of the empire had Shafi'i, Maliki, and Hanbali judges and muftis.

[30] On Ottoman medical pluralism, see Miri Shefer-Mossenshohn, *Ottoman Medicine: Healing and Medical Institutions, 1500–1700* (Albany: State University of New York Press, 2009), especially 22–29.

sickness, and hygiene. We have relatively little information about the third medical tradition, Arab folk medicine, which was largely transmitted orally, but we do know that it relied on remedies based on plants, animals, and minerals, as well as on amulets and charms.

Of the three traditions, this study draws primarily on humoral medicine because it was the most widely practiced in Arab urban centers, including in hospitals, because an abundance of Arabic texts dating from the Ottoman period are available to historians (little has survived of Prophetic or folk medicine), and above all because it was the medical framework with which many Arabs conceptualized impairments.[31] A significant limitation of this body of sources, however, is that they offer little or no information about chronic impairments or physical anomalies, like congenital deafness, that Arab physicians deemed either incurable or not in need of a cure.

The authors of the works I have examined belonged to different socioeconomic classes, had different levels of education, and resided in both major urban centers and smaller, more provincial locales, but they shared three characteristics. First, all were Sunni Muslims. Sunni Muslims represented the vast majority of the population and thus dominated cultural production. Of course, Christian, Jewish, Druze, Shi'i, and other minorities who resided in the region produced their own vast bodies of writings in a variety of fields, which may well contain discourses about impairments that differ from or even contradict Sunni Muslim discourses. The second feature shared by all the authors is that they wrote in Arabic, albeit with varying levels of proficiency. This is hardly surprising, for the Turkish language was foreign to the majority of Arabs living under Ottoman rule. Furthermore, for centuries Arabic had been *the* language of science, medicine, and the law, and therefore, whereas well-educated Turks were expected to at least know how to read Arabic, the reverse was relatively rare. Third, all the authors were male. Despite my best efforts, I was not able to unearth a single work of jurisprudence, medicine, history, biography, or belles-lettres composed

[31] Unfortunately I was not able to locate any hospital records or doctors' patient case files such as, for example, those of the physician-astrologer Richard Napier, whose more than two thousand descriptions of men and women deemed mad formed the base of Michael MacDonald's classic study of madness in seventeenth-century England. See Michael MacDonald, *Mystical Bedlam: Madness, Anxiety and Healing in Seventeenth-Century England* (Cambridge; New York: Cambridge University Press, 1981).

by a Sunni Muslim woman in the Arab East between the mid-sixteenth and the early nineteenth century. There is ample evidence of female literacy in that period, at least among the upper classes, but if women authored any works, those remain in private collections, were written anonymously or under male pseudonyms, have been improperly cataloged, or are lost.

Do these sources constitute an elite body of knowledge, and if so, does this book merely examine a cultural sphere removed from and irrelevant to a broader social reality? This question is conceptually flawed, because it (at least implicitly) posits an impermeable dichotomy between "elite" and "popular" culture that today lacks purchase with most cultural historians. Despite the relative prestige afforded by their education, most of the authors were not denizens of any "ivory tower," but, rather, were actively engaged in the lives of the communities in which they lived. Besides, writers like the Damascene chroniclers Ibn Ayyub al-Ansari (d. ca. 1592) and Ibn Budayr al-Hallaq (fl. 1762), who left us two of the most engaging and colorful accounts of life in Damascus in the early modern period, were of fairly modest means and displayed relatively little proficiency in high literary Arabic. Ibn Budayr earned his living as a barber, while Ibn Ayyub was a court clerk whose life ambition, an appointment as deputy judge, eluded him until shortly before his death.

Many other types of sources were not examined. For example, I did not make use of the voluminous records of the Islamic courts.[32] Ottoman imperial edicts, censuses, and taxation records for the Arab provinces might contain valuable information about the Ottoman government's policies toward people with impairments. A preliminary reading of the Ottoman tax registers for sixteenth-century Syria and Palestine reveals that, in accordance with the Hanafi school of law, blind, lame, and insane men were exempted from certain taxes on the grounds that they were "unsuitable for work" (*amel-mande*).[33] But

[32] I initially intended to include them but abandoned the enterprise after a sampling of the court records of different Arab towns yielded relatively little new information. Future research, however, may well prove me wrong.

[33] Bernard Lewis, "Studies in the Ottoman Archives – I," *Bulletin of the School of Oriental and African Studies, University of London* 16, no. 3 (1954): 474. Other groups exempted from Ottoman taxation included the descendants of the Prophet Muhammad and people employed in religious occupations such as prayer leader, muezzin, and preacher.

perhaps the most promising source for historians of disability is Arabic poetry. The snippets I have read of this enormous corpus of material, much of which is still unedited, reveal beautifully vivid representations and self-representations of and by people with impairments.

This book merely attempts to lay the foundations for a history of disability in the Ottoman Arab world. I hope that future scholarship will refine and perhaps amend the picture I have drawn through new approaches and new sources.

1

Deafness and Muteness

The Mutes of the Seraglio

European visitors to the Ottoman court marveled at the sight of the sultan conversing with his deaf and mute attendants by means of signs. Consider, for example, the following account by Sir Paul Rycaut, the seventeenth-century English author of the *History of the Present State of the Ottoman Empire*:

> Besides the Pages, there is a sort of Attendants to make up the *Ottoman* Court, called *Bizebani* or *Mutes*; men naturally born deaf, and so consequently for want of receiving the sound of words are dumb [. . .] In the day time [. . .] they learn and perfect themselves in the language of the *Mutes*, which is made up of several signs in which by custom they can discourse and fully express themselves; not only to signifie their sense in familiar questions, but to recount Stories, understand the Fables of their own Religion, the Laws and Precepts of the *Alchoran*, the name of *Mahomet*, and what else may be capable of being expressed by the Tongue [. . .] But this language of the *Mutes* is so much in fashion in the *Ottoman* Court, that none almost but can deliver his sense in it, and is of much use to those who attend the Presence of the Grand Signior, before whom it is not reverent or seemly so much as to whisper.[1]

It appears that there existed at the Ottoman court a well-developed system of signs that had been transmitted from generation to

[1] Sir Paul Rycaut (d. 1700), *The History of the Present State of the Ottoman Empire* (New York: Arno Press, 1971), 34–35. First published in 1668. The italics appear in the original.

generation of congenital mutes, and that many hearing men and women at court, including the sultan, were able to communicate through signs. Estimates of the number of Palace mutes, called *bizeban* or *dilsiz* (both terms literally mean "without tongue") vary from thirty to as many as one hundred.[2] There were also mute female attendants and mute concubines at court, but we possess very little information about them.

Like dwarfs and eunuchs, mutes were privileged servants at the Ottoman court. Their tasks included entertaining the sultan, serving at confidential meetings, delivering messages, and, at least until the late seventeenth century, acting as his personal executioners, strangling the condemned with bowstrings. By virtue of their close association with the ruler, Palace mutes could acquire much power and influence. This power occasioned envy, resentment, and disdain, and at times brought on them the accusation of interfering in the affairs of the state.[3]

The question of how and why deaf-mute servants made their way to the Imperial Palace remains to be investigated, but there is evidence that they began to be employed in the fifteenth century, as the empire was expanding. They apparently belonged to different ethnic groups, which suggests that a recruitment system may have been in place. Throughout history people with various mental and physical impairments, especially dwarfs, were kept as jesters by royal courts worldwide, but the practice of keeping mutes was unique to the Ottoman court. We do not yet have a conclusive explanation for this peculiarity, but it may have served, at least in part, a symbolic function: the perfection of the sultan was highlighted through the contrast with the

[2] Thomas Dallam (d. after 1614), *Diary for 1599: Account of an Organ Carryed to the Grand Seignor and Other Curious Matter. Early Voyages and Travels in the Levant. I – The Diary of Master Thomas Dallam, 1599–1600*, ed. J. Theodore Bent (London: Hakluyt Society, 1893), 69–70.

[3] Women, eunuchs, and dwarfs were also viewed with suspicion in this regard. Dikici, "Imperfect Bodies," 77. There is hardly any scholarship on this topic besides Dikici's Master's thesis and M. Miles's 2000 pioneering article "Signing in the Seraglio." See also Miles's valuable annotated bibliography of primary and secondary sources titled "Deaf People, Sign Language and Communication, in Ottoman and Modern Turkey: Observations and Excerpts from 1300 to 2009. From Sources in English, French, German, Greek, Italian, Latin and Turkish, with Introduction and Some Annotation," 2009. Internet publication. URL: http://www.independentliving.org/miles200907.html.

physical imperfection of the deaf-mutes, the dwarfs, and the eunuchs. The preference for congenital mutes over other "imperfects," such as blind men, may also be attributable to the fact that silence was deemed to befit the semidivine status of the sultan.[4]

Whatever the reasons for the mutes' employment at the court, there is no doubt that their presence astonished European visitors. Some Europeans expressed genuine admiration for the mutes' expressive abilities, but their reactions were more frequently tinged with suspicion and contempt. A Venetian diplomat in the mid-sixteenth century, for example, was stunned that Sultan Suleyman would spend so much time exclusively in the company of "eunuchs, mutes, and other most abject specimens of men who are his slaves [. . .]."[5]

An obvious explanation for these responses is that Europeans did not have previous knowledge of a sign system as elaborate as the one in use at the Ottoman court and that they were not accustomed to the sight of deaf courtiers wielding power and eliciting respect. Some of their amazement must also have been rooted in a historically constructed conception of muteness as old as the works of Aristotle that equated the inability to hear and speak with stupidity. Another foreign visitor to the Topkapı Palace, Adolphus Slade, left a vivid account of his encounter with two deaf-mute boys. Slade's prejudiced notions and expectations of brutishness were evidently undermined when he was confronted with the boys' wit and intelligence:

> We delayed a few minutes to converse with two regular mutes; they were boys about 14 years old, very genteel and good-looking, whereby we were completely undeceived in regard of their species, having previously understood that a mute was a kind of animal between a dwarf and a monkey. The little urchins were exceedingly amused, and laughed and conversed about us with great rapidity, making most expressive language about us with their eyes and fingers. Their quick wit is proverbial in Turkey, and in the secret deliberations

4 Dikici, "Imperfect Bodies," iv, 1, 52–54, 63–68. See also Maria-Pia Pedani, "Il silenzio del sultano," in *Il potere della parola, la parola del potere. Tra Europa e mondo arabo-ottomano, tra Medioevo ed età moderna*, ed. Antonella Ghersetti (Venice: Filippi Editore, 2010), 99–112.

5 Antonio Erizzo, "Sommario della relazione di Antonio Erizzo bailo a Costantinopoli, letta in Senato nel 1557," in *Relazioni degli ambasciatori veneti al Senato durante il secolo decimosesto*, ed. Eugenio Albèri (Firenze: Società editrice fiorentina, 1855), vol. 9 (series 3, vol. 3), 138. This encounter is also mentioned in Miles, "Deaf People," 32–33.

of the seraglio, where they alone are allowed to be present as domestics, nothing escapes their intelligence.[6]

This remarkable aggregation of deaf-mute courtiers appears to have been restricted to the Ottoman Palace, but we do know that at least some deaf-mute attendants were also employed in the households of notables in the Arab provinces of the empire. This chapter explores the underlying ideas about deafness and muteness that informed how deaf-mute courtiers were viewed and the extent to which deaf and deaf-mute people were perceived as religiously, spiritually, or intellectually deficient.

Deaf and Mute: A Note on Terminology

In the contemporary United States the word "deaf" in the lower case usually refers to both prelingual and postlingual deafness. Prelingual deafness encompasses congenital deafness, acquired through heredity or pregnancy-related illness, and deafness occurring after birth but prior to the acquisition of a first language. Until fairly recently the terms most commonly used to indicate the prelingually deaf were "deaf-mute" or "deaf and dumb." Today in English the designation "deaf and dumb" is considered offensive because it posits a relationship between the inability to speak and deficient intelligence or "dumbness." Usage of "deaf-mute" and "mute" has also declined because both terms are technically incorrect, since most deaf people have functional vocal cords but have difficulty modulating their voices because they cannot hear themselves or others speak. "Mute" also presumes, incorrectly, incompetence in language and communication.

In the United States, the uppercase "Deaf" has come to signify an identity, as it refers to a specific population of deaf people who share a common language, American Sign Language (ASL), and a common

6 Adolphus Slade, *Records of Travels in Turkey, Greece, etc., and of a Cruise in the Black Sea with the Capitan Pasha in the Years 1829, 1830 and 1831* (Philadelphia and Baltimore: Carey and Hart, 1833), 1:239. Also cited in Nicholas Mirzoeff, "Framed: The Deaf in the Harem," in *Deviant Bodies: Critical Perspectives on Difference in Science and Popular Culture*, ed. Jennifer Terry and Jaqueline Urla (Bloomington: Indiana University Press, 1995), 62.

culture.[7] In this sense, it is comparable to the term "Black." In fact, deafness has a conflicted relationship with disability, as some deaf people who speak ASL do not consider themselves disabled, but rather members of a cultural and linguistic minority.

Postlingual or adventitious deafness, on the other hand, refers to the partial or complete loss of hearing as a result of injury, disease, or advanced age, after the acquisition of language. Consequently, people who are postlingually deaf are usually able to speak. Other terms currently in use are "hard of hearing" and "hearing impaired" (the latter deemed pejorative by some because of its emphasis on impairment), both of which can denote mild, moderate, or severe hearing loss. People who are "hard of hearing" generally use speech as their primary mode of communication but may also be involved in the U.S. Deaf community.[8] For the sake of clarity I have chosen to employ "deaf-mute," "deaf and mute," and "mute" throughout this chapter when discussing prelingual deafness, and "deaf" when referring to postlingual deafness.

In the Ottoman Arab world three Arabic words were commonly used to refer to deafness and muteness: *ṣamam* (masc. adj. *aṣamm*), *ṭarash* (masc. adj. *aṭrash*), and *kharas* (masc. adj. *akhras*). According to a renowned eighteenth-century Arabic lexicon, *ṣamam* is suffering from "obstructions of the ear and the dullness of hearing." *Ṭarash* was generally a synonym for *ṣamam*, although some used *ṭarash* to specify a milder form of *ṣamam*. Both *ṭarash* and *ṣamam* can indicate a broad spectrum of conditions, from mild to severe to complete hearing loss, and the precise meaning can only be deduced contextually. The author of the lexicon notes that the term *ṭarash* was very common in popular parlance, as is the case to this day.

The third word, *kharas* (muteness), means either prelingual deafness or, less commonly, the postlingual temporary or permanent loss of the capacity for speech. A fourth Arabic term used to describe

7 This discussion applies only to the U.S. context. Deaf people in other countries speak different sign languages (for example, Mexican deaf people use Mexican Sign Language and French people French Sign Language).

8 For a comprehensive explanation of all these terms, see the website of the National Association of the Deaf at http://www.nad.org/issues/american-sign-language/community-and-culture-faq. Accessed July 23, 2013.

muteness is *bakam* (masc. adj. *abkam*), which may refer to someone who cannot speak or to someone who cannot speak, hear, or see from birth, or to a mute person who is also affected by idiocy.[9] I have, however, only rarely encountered this word in the legal, medical, and biographical literature, possibly because *bakam* was more commonly used metaphorically rather than to describe the physiological inability to speak.

Prevalence, Causes, and Remedies

There is no statistical data that would allow us to quantify the prevalence of deafness and muteness in the early modern Arab-Islamic world, but we can safely assume that many communities would have included deaf and mute individuals. The rate of hereditary deafness is much higher in regions that practice consanguineous marriages, and first-cousin marriages have been widely practiced throughout the Middle East and North Africa well into the present day. A study conducted among Palestinians, for example, indicates that the prevalence of congenital deafness in 2002 was approximately 0.7 percent, one of the highest in the world (the world average has been estimated at around 0.1 percent).[10] In certain communities the rate is even higher: recent ethnographies of the Al-Sayyid Bedouin community, who live in Israel's Negev desert and among whom consanguineous marriages are the norm, report that 3 percent of the population (about 100 out of 3,000) are deaf from birth.[11] Accidents, pregnancy-related illnesses like rubella, and nutritional deficiencies such as iodine

[9] Muhammad Murtada al-Zabidi (d. 1791), *Taj al-'Arus* (Benghazi, Libya: Dar Libya li-l-Nashr wa-l-Tawzi', 1966), 4:136–137, 4:319, 8:204–205, 8:368. See also Radi al-Din Muhammad Ibn al-Hanbali (d. 1563), *Bahr al-'Awamm fi-ma Asaba fi-hi al-'Awamm*, ed. Sha'ban Salah (Cairo: Dar al-Thaqafa al-'Arabiyya, 1990), 255. *Bahr al-'Awamm* is a tract on colloquialisms in use during the lifetime of the Aleppine Ibn al-Hanbali.

[10] Hashem Shahin et al., "Genetics of Congenital Deafness in the Palestinian Population: Multiple Connexin 26 Alleles with Shared Origins in the Middle East," *Human Genetics* 110 (2002): 284.

[11] Aviad E. Raz and Marcela Atar, "Upright Generations of the Future: Tradition and Medicalization in Community Genetics," *Journal of Contemporary Ethnography* 33, no. 3 (2004): 303. See also Shifra Kisch, "Negotiating (Genetic) Deafness in a Bedouin Community," in *Genetics, Disability, and Deafness*, ed. John Vickery Van Cleve (Washington, DC: Gallaudet University Press, 2004), 148–173.

deficiency would also have resulted in the birth of deaf-mute children. Finally, residents of the Ottoman Arab East, like everywhere else, could have lost their hearing through a host of childhood and adult illnesses, ear infections, injuries, extended or excessive exposure to noise, or through the natural deterioration of the auditory nerves that accompanies aging.

I have not found any discussions of *kharas* (muteness) as a medical condition, but early modern medical texts examine the various forms and causes of deafness in some detail. For example, the famous physician Da'ud al-Antaki (d. 1599), who was based in Egypt, devoted sections of his books to the anatomy and diseases of the ear. Al-Antaki explains that although the words *ṣamam* and *ṭarash* are used interchangeably in popular parlance to describe the partial or complete lack of hearing, they actually refer to different conditions. To be precise, *ṣamam* describes the total absence of hearing and is often congenital, in which case it results in the inability to speak. *Ṭarash*, on the other hand, tends to refer to the deficiency or diminution of the faculty of hearing and is usually accidental. As for remedies, al-Antaki asserts that there is rarely a cure for *ṣamam*. *Ṭarash*, however, can, depending on its etiology, be healed or at least relieved. If caused by old age, no remedy exists. If caused by an obstruction in the ear, then the obvious course of action is to try to remove the blockage. If it is the result of a blow, then it is necessary to identify and treat the affected nerve. If the root of the problem is a humoral imbalance, equilibrium must be restored by performing a phlebotomy or with one of al-Antaki's own many "tried and tested" medicinal recipes.[12]

[12] Da'ud al-Antaki (d. 1599), *Tadhkirat Uli al-Albab wa-l-Jami' li-l-'Ajab al-'Ujab* (Beirut: al-Maktaba al-Thaqafiyya, n.d.), 1:16; 2:18–19, 153; 3:81–82, 142–143. See also al-Antaki, *al-Nuzha al-Mubhija fi Tashhidh al-Adhhan wa-Ta'dil al-Amzija*, printed on the margins of his *Tadhkirat Uli al-Albab*, 1:126–127, 309; 3:16–21, and Ahmad al-Qalyubi (d. 1659), *al-Tadhkira fi al-Tibb*, eds. Ahmad Farid al-Mazidi and Muna Shalabi with the title *Tadhkirat al-Qalyubi fi al-Tibb wa-l-Hikma* (Beirut: Dar al-Kutub al-'Ilmiyya, 2001), 43–45. A detailed biography of Da'ud al-Antaki can be found in Darwish al-Talawi's (d. 1606) *Sanihat Duma al-Qasr fi Mutarahat Bani al-'Asr*, ed. Muhammad Musa al-Khawli (Beirut: 'Alam al-Kutub, 1983), 2:32–52. Al-Talawi, a famous poet, teacher, and mufti from Damascus, met the renowned physician during one of his travels and composed a biographical sketch of him on the basis of their conversations. The life and work of al-Antaki, who authored dozens of tracts and continued to be cited as a major medical authority well into the nineteenth century, has received surprisingly little attention from modern scholars.

The renowned physician Ibn Sallum (d. 1670/1) also dedicated a section of his medical treatise to various ailments of the ear. He provides the following description of deafness: "It is of two types: either severe, in which there is no hearing, or mild, in which hearing is possible. It can be either in the ear itself or in the nerves, if an obstruction occurs in them. The cause of deafness is most frequently thick phlegm that is implanted in the ear or nerves."[13] Ibn Sallum, like al-Antaki, states that *ṭarash* may be caused by a blow, a fall, or, citing Hippocrates, a corrupt humor (the choleric humor in particular). In the latter case, phlebotomy, induced vomiting, or the administration of a purgative may prove helpful. Ibn Sallum adds that several other "malignant illnesses," such as epilepsy, leprosy, and "the Frankish pox" (syphilis) can cause deafness. Lastly, deafness may appear in the aftermath of hemiplegia (partial paralysis).[14]

None of the medical works I have examined offer first-hand accounts or case studies of specific women or men affected by deafness (or by any other disease, for that matter). We are fortunate, however, to have the medical history of a young woman in Tripoli in present-day Lebanon recorded in some detail by a European physician by the name

For one recent biographical overview of al-Antaki, see Julia Bray, "Dawud ibn ʿUmar al-Antaki," in *Essays in Arabic Literary Biography, 1350–1850*, ed. Joseph E. Lowry and Devin J. Stewart (Wiesbaden, Germany: Harrassowitz Verlag, 2009), 2:48–56.

13 Salih b. Nasr Allah al-Halabi, known as Ibn Sallum, *Ghayat al-Itqan fi Tadbir Badan al-Insan*, MS, Dar al-Kutub al-Wataniyya, Cairo, al-Madrasa al-Ahmadiyya 1044, 52b. We possess scant information on the life of Salih b. Nasr Allah al-Halabi, better known as Ibn Sallum or Salih Efendi. We know that he was born and educated in Aleppo, where he practiced medicine and served as chief of physicians. At some point in his career he moved to Istanbul, where Sultan Mehmet IV (r. 1648–1687) appointed him Chief Physician. There is some evidence that he also received judicial appointments, and that he died in Yenişehir in 1670 or 1671. Although not nearly as prolific an author as Da'ud al-Antaki, Ibn Sallum's contributions to Ottoman medical knowledge were equally if not more significant, for he is credited with introducing early modern European Paracelsian "chemical medicine" into Islamic medicine. See Muhammad Amin al-Muhibbi (d. 1699), *Khulasat al-Athar fi Aʿyan al-Qarn al-Hadi ʿAshar* (Beirut: Dar Sadir, 1966), 2:240–242. See also Max Meyerhof and M. Monnerot-Dumaine, "Quelques maladies d'Europe dans une encyclopédie médicale arabe du XVIIe siècle," *Bulletin de l'Institut d'Égypte* 24 (1942): 33–47; Esin Kahya and Ayşegül Demirhan Erdemir, *Medicine in the Ottoman Empire (And Other Scientific Developments)* (Istanbul: Nobel Medical Publications, 1997), 81–85; Abdülhak Adnan Adivar, *La science chez les Turcs ottomans* (Paris: G.-P. Maisonneuve, 1939), 96–98, and Shefer-Mossensohn, *Ottoman Medicine*, 176–178.

14 Ibn Sallum, *Ghayat al-Itqan*, 52b.

of Raymond de Vermale, who spent much of the period between 1723 to 1737 in the region. He describes the onset of deafness in his patient in the following terms:

> On the subject of ear hemorrhages, I have seen a girl who, at the age of nineteen, despite having very regular menstrual periods, began to suffer for two or three years from a ringing in her ears that intensified immediately prior to each period. This ringing was accompanied by headaches that extended to the jugulars and lasted until the end of her menses. One night, while preparing for bed with one of her friends, she burst into such a fulsome fit of laughter that she felt a small rupture inside her ears. The next day, the bed was, so to speak, inundated with blood. From that moment, the blood flowed moderately for nine consecutive months, ceasing from the 15th of July to the 15th of October 1733, when this small hemorrhage began anew. Despite the consistent regularity of the patient's menses, this hemorrhage reoccurred periodically for five years, with the aforementioned occasional variation, that only a bit of purulent matter (but more often only blood) would emerge. This would stop regularly from around the 15th of July until the 15th or 18th of October [of every year], when it would start again [...]. The patient took great care to record these changes in her almanac, while refusing any treatment whatsoever. This continued until the 21st of August 1737, when a bout of fever convinced her to undergo bloodletting for the first time in her life. This stopped the hemorrhage and left her with deafness, but she hears much less when one shouts in her ears than when one speaks with moderate volume to her face.[15]

This is a rare glimpse both of the process by which adventitious deafness was acquired and of how physicians treated it. Vermale, whose European education was of the same medical tradition as that of Ottoman physicians, that is, one steeped in humoral theory, also recommended bloodletting. Vermale's repeated mention of the young woman's menstrual periods conformed to this theory, since physicians understood physiological functions like menstruation in that general context. All human beings experienced, with more or less frequency

[15] The original French text appears in Pierre Huard, "Castration judiciaire en 1726 à Tripoli (de Syrie)," *Clio Medica* 2 (1967): 368–369. The title refers to another medical history transcribed by Huard, that of a man castrated by order of the Pasha of Tripoli for raping a young virgin. The sparse biographical information available on Raymond de Vermale is collected in Claude Marie Dordain's *Les membres non résidents étrangers de l'Académie Royale de Chirurgie*, Thèse de Rennes no. 397 (1964). Vermale, who died sometime after 1767, was admitted as a member of the Académie Royale de Chirurgie in 1739. He specialized in surgery and published several articles in the *Journal de Médecine, de Chirurgie et de Pharmacie*.

and regularity, a surfeit of blood that could be discharged from the body through a number of means and via a variety of routes. The cyclical discharge of blood that characterized menstruation, not the mere fact of the discharge, was seen as a physiological process distinctive to the female.

Beyond the realm of medicine and humoral theory, there existed other competing explanations for the onset of deafness and muteness. One incident reported by Ibn Tulun, a chronicler of early Ottoman Syria and Egypt, suggests that deafness and muteness were sometimes attributed to supernatural agency. Ibn Tulun writes, without context or commentary, that one day in Damascus a jinni (an invisible supernatural being) struck a man while he was urinating in a latrine. So powerful was the blow that its recipient permanently lost the power of speech.[16] Unfortunately, alternative explanations of this kind were rarely recorded in the early modern period, and the vast biographical literature of the early modern Arab-Islamic world indicates that people lost their hearing for far more mundane and unglamorous reasons, including old age, accidents, and illnesses. It is to the lives of these individuals that we now turn.

Deafness as Lived Experience

It was not unusual to lose one's hearing in the Ottoman Arab world. Siraj al-Din al-Maqdisi (d. 1594/5), a teacher and mufti in Jerusalem, lost his hearing as a result of old age.[17] Others became partially or completely deaf as a result of accidents or illnesses. For example, Ibn al-Furfur (d. 1627) was born into a prominent Damascene family of judges and scholars and was able to receive a first-rate education under the guidance of several acclaimed teachers. He excelled particularly in Hanafi law, which he taught in a local school. In the midst of a successful life, his biographer recounts, "it happened that fate struck his auditory meatuses with the plug of deafness." Following the onset of

[16] Shams al-Din Muhammad Ibn Tulun (d. 1546), *Mufakahat al-Khillan fi Hawadith al-Zaman: Tarikh Misr wa-l-Sham*, ed. Muhammad Mustafa (Cairo: al-Mu'assasa al-Misriyya al-'Amma li-l-Ta'lif wa-l-Tarjama, 1962–1964), 1:259.

[17] Al-Muhibbi, *Khulasat al-Athar*, 3:220–221. For other examples of hearing loss because of advanced age, see the biographies of Ahmad al-Ghanimi and Muhammad al-Muhibbi, ibid., 1:312–315, and 4:231–233, respectively.

deafness, Ibn al-Furfur reduced his participation in public life, congregating only with a few intimates. He chose to occupy himself with the family business and became the administrator of his family's *waqf* (pious endowment). He also developed a marked interest in composing and solving riddles, for which he became renowned.[18]

Like Ibn al-Furfur, 'Ali al-Shami lost much of his hearing in adulthood. The son of a prominent mystic, 'Ali was born in the vicinity of Beirut in 1501. He began studying the Qur'an at the age of five and memorized it in the span of two years. 'Ali then studied with several illustrious scholars of the time and became well versed in a variety of disciplines, including Hadith, law, and arithmetic. Described as urbane and dignified, he was also deemed an excellent poet by his peers. In his late teens he settled in Medina with his family. 'Ali appears to have lost most or all of his hearing during his residence in Medina, because old acquaintances who met with him in 1546 during one of his frequent journeys to Syria remarked that "he was unlucky with his hearing, for severe deafness had come upon him." According to his father, 'Ali's deafness could be traced directly to the fact that he was "addicted to listening to beautiful tunes." Not heeding his father's exhortations to abandon this trivial habit, "he was punished" with the loss of his hearing. His deafness notwithstanding, 'Ali was appointed imam (prayer leader) and preacher of the Mosque of the Prophet Muhammad in Medina.[19]

It was possible for men who were hard of hearing to receive high-level judicial appointments. Muhammad b. Da'ud, known as Riyadi *al-uṭrūsh* (the deaf, a variation of *aṭrash*) al-Rumi, arrived in Damascus in 1617 to take up the prestigious post of chief judge of the city. Riyadi's biographer describes him as an unpopular and greedy judge whose wife was known for devoting her time to "fun and games." His wife's apparent frivolity, in conjunction with his very poor hearing, became the subject of humorous anecdotes told at the judge's expense.

[18] Ibn al-Furfur's biography is recorded in al-Muhibbi, *Khulasat al-Athar*, 1:299–301.

[19] Najm al-Din Muhammad al-Ghazzi's *al-Kawakib al-Sa'ira bi-A'yan al-Mi'a al-'Ashira*, ed. Jibra'il Sulayman Jabbur (Beirut: Manshurat Dar al-Afaq al-Jadida, 1979), 2:197–199, and Radi al-Din Muhammad Ibn al-Hanbali (d. 1563), *Durr al-Habab fi Tarikh A'yan Halab*, ed. Mahmud Ahmad al-Fakhuri and Yahya Zakariya 'Abbara (Damascus: Manshurat Wizarat al-Thaqafa, 1972–1974), 1:1004–1010.

For example, one night, on hearing the sound of musical instruments coming from his wife's quarters, Riyadi demanded to know what was happening. When Riyadi's wife boldly replied, "It is the muezzins calling to prayer!" he reputedly believed her.[20]

The biographies of 'Ali al-Shami, Ibn al-Furfur, and Riyadi illustrate a critical point: adventitious deafness was not necessarily a social liability, at least not for those who were living in relatively privileged circumstances prior to suffering hearing loss. Nor, apparently, did it necessarily impede the pursuit of a relatively normal life. Riyadi held the plum post of chief judge of Damascus. 'Ali continued to study, write, travel, and maintain an active social life and regular contact with other scholars. His contemporaries even credited him with spreading among the people of Damascus the novel habit of drinking coffee during one of his extended visits to Syria in the 1540s.[21] Perhaps most telling was his appointment as the imam of one of the most important mosques in the Muslim world. And although Ibn al-Furfur chose to lead a more secluded life, his impairment did not prevent him from either earning a living or pursuing his intellectual interests.

Ultimately, how one responded to deafness depended on the degree of hearing loss as well as on the personal circumstances and attitudes of the affected individual. For some it was a source of terrible distress. 'Abd Allah al-Yusufi al-Halabi (d. 1780) was an accomplished and prolific poet from Aleppo. He earned his living selling coffee beans, and for a period of time also delivered sermons in a local mosque. According to his biographer, he lived in poverty and misery. Three years before his death he was struck by severe deafness, which worsened to the point that he was entirely unable to hear and people started communicating with him through signs (*bi-l-ishāra*). Deafness was such a source of hardship for him that he vowed to compose one thousand verses dedicated to the Prophet Muhammad in the hopes of being healed.[22]

20 Al-Muhibbi, *Khulasat al-Athar*, 3:463–464. Also in Najm al-Din Muhammad al-Ghazzi (d. 1651), *Lutf al-Samar wa-Qatf al-Thamar min Tarajim A'yan al-Tabaqa al-Ula min al-Qarn al-Hadi 'Ashar*, ed. Mahmud al-Shaykh (Damascus: Manshurat Wizarat al-Thaqafa wa-l-Irshad al-Qawmi, 1981–1982), 1:124–125.

21 Al-Ghazzi, *al-Kawakib al-Sa'ira*, 2:198.

22 Muhammad Khalil al-Muradi (d. 1791), *Silk al-Durar fi A'yan al-Qarn al-Thani 'Ashar*, ed. Akram Hasan al-'Ulabi (Beirut: Dar Sadir, 2001), 3:121–129. For another

Others responded with less distress to the onset of deafness. Mahmud b. al-Bayluni, the son of a renowned scholar, studied in Aleppo with the luminaries of his time and subsequently became an accomplished scholar and teacher himself. In January of 1599, just a few months prior to his death, al-Bayluni entered Damascus, a way station on his pilgrimage to Mecca. While in the city he spent some time in the company of local scholars, including his future biographer, Najm al-Din al-Ghazzi, who was most impressed by his erudition. Al-Ghazzi reports that al-Bayluni, who was over sixty, was so deaf that he could not comprehend speech unless it was delivered at considerable volume directly into his ears. Yet he expressed a positive approach to his condition: "This deafness is a gift from God," he explained to al-Ghazzi, because he was no longer distracted by the gossip and frivolous chattering of humans but could choose instead to hear only the Qur'an, when recited close to him.[23]

These and other biographies of deaf men show that moderate or severe hearing loss was generally viewed as a misfortune, but it did not inevitably constitute an insuperable impediment to participation in social and economic life. The inability to speak, on the other hand, was by all indications just such an impediment. The biographical dictionaries are silent about the lives of the prelingually deaf. In fact, I found only one reference to a man explicitly described as both deaf and mute (although we cannot be certain that he was born deaf) in the thousands of biographical entries I have consulted. The account is so perfunctory that it is impossible to draw any meaningful conclusions, but it provides anecdotal evidence that being deaf and mute was viewed with some degree of compassion. In 1612 a landowner by the name of 'Abd al-Ghani b. al-Dawidar traveled to a village in Wadi al-Taym (in present-day Lebanon) to take care of some business. He spent the night in the house of an unnamed deaf and mute man (*akhras aṭrash*) from that village. When he got up in the middle of the night to

example of the distress that deafness could cause, see the brief biography of 'Abd al-Ghani al-Salihi (d. 1772) in ibid., 3:45.

[23] Al-Ghazzi, *Lutf al-Samar*, 2:628–637. For more biographies of deaf men, see the Egyptian Ahmad b. Muhammad al-Bani, who was deaf (*aṣamm*) like his father, in al-Ghazzi, *al-Kawakib al-Sa'ira*, 1:129; of the Damascene Ahmad b. Hajji *al-uṭrūsh* in ibid., 1:133; and of the Egyptian Ahmad al-Subki in al-Muhibbi, *Khulasat al-Athar*, 1:185–186.

answer a call of nature, he stumbled over the deaf man, who mistook him for an assailant and stabbed him with a dagger. Al-Dawidar died of his wounds four days later. Before dying, he instructed his son not to pursue legal action against his killer, presumably because he realized that the deaf-mute man had acted instinctively out of panic and confusion.[24]

The absence of biographies of deaf and mute men is not surprising given that the world depicted in the biographical literature is largely that of learned men who acquired and demonstrated their erudition in a milieu that placed a premium on verbal skills. For someone born deaf, acquiring the requisite skills for admission to this community would have entailed extraordinary challenges. Just learning to read and write would have constituted a Herculean task. As will become clearer in the next chapter, orality was paramount in the world of premodern education and learning. If blind people were not necessarily disadvantaged in this respect, deaf-mutes most certainly were. This was apparent to contemporary scholars. In a section of a treatise dedicated to blindness, ʿAli al-Qari (d. 1606) presented a debate among the learned on the relative preferability of blindness over deafness. According to the author, the parties to this debate agreed on the undeniable fact that many prominent and respected scholars were born blind, but the same could not be said of persons born deaf.[25]

Does God Hear Silent Prayers? The Mute and Religious Ritual

Even beyond the relatively narrow world of the learned, a person who could neither speak nor hear would have been at an enormous disadvantage in oral societies like those of Ottoman Egypt and Syria, where everyday business relied on verbal agreements.[26] But how was this disadvantage reflected in the law? Islamic law did not impose

[24] Hasan al-Burini (d. 1615), *Tarajim al-Aʿyan min Abnaʾ al-Zaman*, ed. Salah al-Din al-Munajjid (Damascus: Matbuʿat al-Majmaʿ al-ʿIlmi al-ʿArabi bi-Dimashq, 1959 and 1963), 2:370.

[25] ʿAli b. Sultan Muhammad al-Harawi al-Qari (d. 1606). *Tasliyat al-Aʿma ʿala Baliyat al-ʿAma*, ed. Mabruk Ismaʿil Mabruk (Cairo: Maktabat al-Qurʾan, 1994), 27–28.

[26] James Grehan has aptly illustrated how in Ottoman Damascus, for example "most people relied heavily on verbal pacts to conduct business, conclude agreements, and enforce contracts. [...] In many ways, people were only as good as their words." James Grehan, "The Mysterious Power of Words: Language, Law, and Culture in

restrictions on those who experienced hearing loss; rather, it strove to acknowledge the realities confronting deaf people and to accommodate their needs. Muhammad al-Tumurtashi, a sixteenth-century Hanafi mufti from Gaza and the author of an influential law compendium, was one day asked to issue a fatwa concerning the validity of a divorce between a woman and her deaf husband.

Question: About a deaf (*uṭrūsh*) man who divorced his wife with a revocable divorce [i.e., he did not proclaim the divorce formula three times]. He owes her three dinars for her dower. She brought a suit against him to the judge. The judge said to the man: "Did you divorce her three times?" He replied "Yes," believing that the judge was inquiring about the woman's abovementioned dower. If that is his claim and he denies that he understood the judge's question about the three divorces, is such a statement accepted, given that his deafness is known among the people, or not?

Answer: Yes, his statement about that [the misunderstanding] is accepted if given under oath, because of the appearance of the evidence attesting to his truthfulness and because he denies that the divorce has taken place [. . .].[27]

The mufti's response reflects Islamic law's general spirit of pragmatism and its effort to strike a balance between the interests of the individual and those of the community.

This spirit is also reflected in the law's provisions for the *akhras*, a legal category that encompassed two types of mutes: persons born deaf and therefore incapable of intelligible speech and those who become mute after learning to speak, as a result of injury or illness. It is difficult to discern with certainty when *akhras* refers to deaf-mutes or just to mutes in the legal literature. However, since the jurists themselves pointed out that the second category, those who were mute but not deaf, was far less common, we may assume that most of the discussions of *akhras* refer to deaf-mutes.[28]

Ottoman Damascus (17th–18th Centuries)," *Journal of Social History* 37, no. 4 (2004): 991–992.

27 Muhammad b. 'Abd Allah al-Ghazzi al-Tumurtashi (d. 1596), *Fatawa*, MS, Princeton University Library, Yahuda 4135, 23b. For another example, see Hasan al-Husayni (d. 1811), *Fatawa*, MS, Princeton University Library, Yahuda 5415, 178a.

28 Muhammad Amin Ibn 'Abidin (d. 1836), *Radd al-Muhtar 'ala al-Durr al-Mukhtar Sharh Tanwir al-Absar*, ed. 'Adil Ahmad 'Abd al-Mawjud and 'Ali Muhammad Mu'awwad, 2nd ed. (Beirut: Dar al-Kutub al-'Ilmiyya, 2003), 4:448; 5:136, 160.

In the realm of religious practice, the problem that most concerned Muslim jurists was the mute's inability to verbalize prayer. Although the mute's right to prayer itself was not questioned in Islamic law, there was disagreement over the necessity of moving the tongue and lips to simulate verbalization. On this issue, the Hanafi and Shafi'i schools reached different conclusions. For the Shafi'is, a mute ought to move his tongue during prayer if physically capable of doing so. According to the majority Hanafi opinion, on the other hand, someone who could not speak was not required to move his lips or tongue when praying, because an important component of prayer is *intent*. Prayer in the heart, so to speak, was sufficient.[29]

A more problematic question discussed by legal scholars is whether it is permissible for a mute to lead those who can speak in prayer, that is, to act as imam. When Muslims congregate for communal prayer they are advised to choose as prayer leader the person who is most pious and possesses the most comprehensive knowledge of the Qur'an. All four schools of law agreed that it is not permissible for a mute to lead others in prayer, even if those who speak are all *ummī* – defined variously in Hanafi law as those "who do not have a verse of the Qur'an memorized," or those who are so ignorant of proper ritual that they are not "capable of uttering the *taḥrīma*" (the *takbīrat al-iḥrām*, that is, to pronounce "Allahu Akbar" at the beginning of a prayer). The status of the mute, who is incapable of intelligibly uttering even this simple phrase, is, for the purpose of such rituals, inferior to that of the *ummī*.[30]

Similar questions arose in reference to people with some speech impediments, including lisping, stuttering, and difficulty with pronouncing the letter "r." Speech impediments were explained by physicians as problems of the tongue that might be caused by a humoral imbalance or might be the result of hemiplegia, the paralysis of one side

[29] Layla 'Abd Allah, *Ahkam al-Akhras fi al-Fiqh al-Islami* (Cairo: Maktabat al-Zahra' li-l-Tab' wa-l-Nashr wa-l-Tawzi', 1997), 69; Ibn 'Abidin, *Radd al-Muhtar*, 2:90–91.

[30] Ibn 'Abidin, *Radd al-Muhtar*, 2:324, 343, and 'Ala' al-Din Muhammad al-Haskafi (d. 1677), *al-Durr al-Mukhtar Sharh Tanwir al-Absar*, with a commentary by Ibn 'Abidin in *Radd al-Muhtar*, 2:324. For the Shafi'is, an *ummī* is someone who "is not capable of reciting the Fatiha." The Fatiha is the opening Sura (chapter) of the Qur'an.

of the body due to a stroke.[31] People with speech impediments sometimes had to endure unwelcome mocking, but the impediments, even when very pronounced, did not necessarily prevent them from achieving professional success.[32] For example, Muhammad al-Baghdadi's (d. 1607) speech was difficult to understand: "The stutter on his tongue was so great that he could hardly pronounce his words," commented one biographer. "He had a stutter on his tongue as if he were mute," echoed another. Despite his stutter, he was widely regarded as a remarkably accomplished scholar in a variety of disciplines, including jurisprudence, Hadith, logic, and Arabic, and was granted several lucrative official teaching and administrative posts in Damascus.[33]

Abu Bakr b. 'Ali (d. 1532/3), too, fared well despite his speech impediment and worked as a court scribe in Aleppo. A biographer reported that he had spent some time in his company and had found him to be polished, decorous, lighthearted, and very pleasant, except that he pronounced the letter "r" defectively. The biographer hastened to add that he "did not fault him" for that, for it had happened to many great personages before him, including the famed tenth-century grammarian Ibn al-Sarraj.[34]

Should someone like Abu Bakr be permitted to lead in prayer those who did not have any speech defects? Khayr al-Din al-Ramli, the influential seventeenth-century Hanafi mufti from Ramla, was one day asked this very question.

Question: An imam is *althagh* and exchanges the *rā'* with the *ghayn*, and when he wants to enunciate "al-Rahman al-Rahim" says "al-Ghahman

[31] Al-Antaki, *al-Nuzha al-Mubhija*, 1:128–29; 3:40. The causes of speech impediments remain unknown in many cases. However, there is some evidence that stuttering has a genetic component.

[32] Sharaf al-Din Musa Ibn Ayyub al-Ansari (d. ca. 1592), *al-Rawd al-'Atir fi-ma Tayassara min Akhbar Ahl al-Qarn al-Sabi' ila Khitam al-Qarn al-'Ashir*, partial edition by Ahmet Halil Güneş with the title *Das Kitāb ar-raud al-'ātir des Ibn Aiyūb: Damaszener Biographien des 10./16. Jahrhunderts, Beschreibung und Edition* (Berlin: Klaus Schwarz Verlag, 1981), 43–44; Ibn al-Hanbali, *Durr al-Habab*, 1:385.

[33] Al-Baghdadi's biography is in al-Burini, *Tarajim al-A'yan*, 3:151b–152b; al-Ghazzi, *Lutf al-Samar*, 1:191–94; al-Muhibbi, *Khulasat al-Athar*, 4:31–33.

[34] Ibn al-Hanbali, *Durr al-Habab*, 1:385. For more examples of men with speech impediments, see Ibn al-Hanbali, *Durr al-Habab*, 1:133; Ibn Ayyub, *al-Rawd al-'Atir*, 43–44; al-Ghazzi, *Lutf al-Samar*, 1:243.

al-Ghahim," and when he wants to enunciate "Rabb," says "Ghabb" instead. Is his leading in prayer those who can properly enunciate void? Is being an imam to the eloquent (*faṣīḥ*) not permitted? Should he be prohibited from leading the eloquent in prayer? And should his leadership in prayer of those like himself be discouraged? Should the ruler prevent him from leading prayer in the congregational mosque?

Al-Ramli began his response by noting that this issue had arisen frequently and that jurists had disagreed about the answer. Some, including luminaries of Hanafi jurisprudence, had argued that a man with a speech impediment could indeed lead in prayer people without speech impediments, but al-Ramli's personal opinion was that it was not permissible.[35] For the same reason that mutes should not be allowed to lead the *ummī* in prayer – that leading the faithful in prayer requires specific verbal skills – people with speech impediments should also be prohibited from acting as imams. Again, the governing criteria are functional in nature, as one of the imam's tasks is the clear and precise articulation of ritually prescribed words and phrases.

But what if a group of people with speech impediments or a group of mutes congregates for communal prayer? Could one of their own serve as imam in that situation? Jurists usually replied in the affirmative, reasoning that, given that all participants would be equally impaired in reciting the Qur'an, they should be permitted to pick one from their midst to lead them.[36] The jurists' stance on this issue is intriguing, because it might have paved the way for the creation of a deaf community who prayed together using signs. To date, however, no documentary evidence has been unearthed, if it indeed exists, proving that deaf-mutes gathered for prayer in specific mosques, houses, or

[35] Kayr al-Din al-Ramli (d. 1671), *al-Fatawa al-Khayriyya li-Naf' al-Barriyya* (Cairo: n.p., 1857), 1:9. See also ibid., 9–10. According to the mufti Isma'il Ibn al-Ha'ik, a man was *althagh* only if he transformed the letter *rā'* into a *ghayn*, not if he could simply not pronounce the *rā'* clearly. If the latter was the case, it was indeed permissible for him to lead prayer. Isma'il Ibn al-Ha'ik (d. 1701/2), *Shifa' al-'Alil bi-Fatawa al-Marhum al-Shaykh Isma'il*, MS, Princeton University Library, Yahuda 5535, 2b.

[36] See Muhammad b. 'Abd Allah al-Ghazzi al-Tumurtashi (d. 1596), *Matn Tanwir al-Absar wa-Jami' al-Bihar* (Cairo: Matba'at al-Taraqqi, 1914), 19; al-Haskafi, *al-Durr al-Mukhtar*, 2:327–329; Ibn 'Abidin, *Radd al-Muhtar*, 2:327–330. On the question of the permissibility of deaf-mutes leading prayer, see also Rispler-Chaim, *Disability in Islamic Law*, 25–26.

other venues, except for the Ottoman palace in Istanbul. Today, on the other hand, sign language interpretations at the Friday congregational prayer are available in mosques throughout the Middle East.[37]

Spiritually Sound

That Islamic law did not exclude mutes from the community of the faithful may seem perfectly reasonable to a twenty-first century reader. When placed in a broader and comparative historical perspective, however, it is remarkable. Other contemporaneous societies had significantly different attitudes toward and concerns about mute people's participation in religious ritual. Susan Plann's research shows that in early modern Spain mutes were usually denied participation in activities central to the expression of the Catholic faith. The Catholic Church, ambivalent toward deaf-mutes at the best of times, appears to have played a negative role in this regard. The Church's thinking about this impairment was influenced by the apostle Paul's well-known declaration that "faith cometh by hearing" and by St. Augustine's pronouncements that deafness "hinders faith itself" and that congenital deafness and blindness were punishments inflicted on newborns by God for their parents' sins. These rhetorical excommunications may have been more metaphorical than literal in intended meaning and may have later been misinterpreted or taken out of context, but it is a fact that they were frequently used to justify the exclusion of deaf-mutes from church membership, the sacraments, the possibility of salvation, and other rites and activities ordinarily available to the faithful.[38]

Early modern England provides another interesting comparison. According to Emily Cockayne, post-Reformation England witnessed a curtailment of both the deaf and the deaf-mutes' ability to participate in Church rituals: "After the Reformation the doctrinal culture of

[37] The Egyptian government, for example, inaugurated a new service for its deaf citizens in July 2000: sign language interpretation at the Friday prayer service held at the historic Sayyida Zeinab Mosque in Cairo. According to Ahmed Younis, who heads the Egyptian Organization for the Rights of the Disabled, the government plans to expand this service to other Egyptian mosques. "Interpreting the Word," *Al-Ahram Weekly*, 6–12 July 2000 (no. 489), also available online at http://weekly.ahram.org.eg/2000/489/eg11.htm. Accessed March 30, 2012.

[38] Susan Plann, *A Silent Minority: Deaf Education in Spain, 1550–1835* (Berkeley: University of California Press, 1997), 17–18.

Protestant England placed an increased value upon hearing in relation to the other senses," because "in the Anglican Church the ability to hear was a paramount skill, necessary to absorb the word of God, which was now uttered in the vernacular tongue."[39] As a result, those who could not hear experienced progressively greater exclusion from religious life.

The concept of "original sin" is rejected by Islamic theology. A connection between deaf-muteness and deficient spirituality, however, can be found in the Qur'an. In their overview of disability in the sacred text, Maysaa Bazna and Tarek Hatab point out that the most common word for "mute," *akhras*, is absent in the Qur'an. Instead, *abkam*, "deaf and mute," appears on six occasions. An analysis of each occurrence leads the authors to conclude that *abkam* is not meant to refer to someone who is physiologically deaf and mute but it "is intended to signify one who is spiritually, ethically, or morally bereft." Similarly, in all the fifteen instances in which *aṣamm*, for "deaf," is mentioned (the other common Arabic word for "deaf," *aṭrash*, never appears in the Qur'an), it is employed exclusively metaphorically, sometimes in conjunction with blindness, to indicate the rejection of God's message.[40]

These metaphorical associations of deafness and muteness with deficiency of religious belief make the occasional appearance in the literature of the Ottoman Arab world, notably in dream interpretation manuals. In 'Abd al-Ghani al-Nabulusi's (d. 1731) oneiromantic treatise, for example, the entries on *ṭarash* (deafness) and *kharas* (muteness) parallel the entry on blindness examined in the next chapter. Much like blindness, *ṭarash* "indicates deficiency in religion, because God has said about the unbelievers 'deaf, mute, blind'." Similarly, the primary meaning of *kharas* in a dream is "corruption in religion."[41] It bears repeating, however, that these metaphorical understandings of deafness and muteness were not incorporated into Sunni legal

39 Emily Cockayne, "Experiences of the Deaf in Early Modern England," *The Historical Journal* 46, no. 3 (2003): 496.

40 Maysaa S. Bazna and Tarek A. Hatab, "Disability in the Qur'an: The Islamic Alternative to Defining, Viewing, and Relating to Disability," *Journal of Religion, Disability and Health* 9, no. 1 (2005): 15–16. See also Geert Jan H. van Gelder, "Hearing and Deafness," in *Encyclopaedia of the Qur'an*, ed. Jane McAuliffe (Leiden: Brill, 2002) 2:405–406.

41 'Abd al-Ghani al-Nabulusi, *Ta'tir al-Anam fi Ta'bir al-Manam*, 2nd ed. (Cairo: Dar Ihya' al-Kutub al-'Arabiyya, 1958–1959), 2:69 and 1:194–195, respectively.

theory or practice, and thus they did not result in significant restrictions for the deaf and/or mute in important aspects of life, such as marriage.

Marriage and Divorce in Islamic Law

The marital bond is central to both the theory and the application of Islamic law. According to the consensus of Islamic jurisprudence, the primary purposes of marriage are the legitimation of sexual relations, the provision of sexual fulfillment, and companionship. With very few exceptions, Muslim men and women were strongly encouraged to marry, and mutes were not excluded from this exhortation. There was, however, one fundamental issue that required the attention of the jurists. Marriage is a binding (yet revocable) contract that usually requires the verbal consent of both the bridegroom and the bride. These two parties or their legally empowered representatives, often a male guardian for youth who have yet to achieve majority, conclude the contract by verbally expressing their consent to its terms in the presence of two or more witnesses. The specific provisions of the contract, such as the amount of the dower owed to the bride, may then be recorded in writing.[42]

If a marriage contract requires verbal consent, what options are available to someone who cannot speak? The four major schools of Sunni jurisprudence concurred that the preferable course of action was for the mute to convey his or her consent in writing, the rationale being that the writing of the mute is equivalent to verbal expression. Written consent, however, was not mandatory, as the jurists were fully cognizant of an obvious reality, the prevalence of illiteracy. Thus they determined that if one of the parties were illiterate the marriage contract would still be valid if the mute could articulate his or her assent through a "clear sign" (*ishāra maʿlūma*).[43] In accordance with

[42] For a comprehensive discussion of Hanafi doctrine on marriage and divorce in Ottoman Syria see chapters 2 and 3 of Judith Tucker's *In the House of the Law: Gender and Islamic Law in Ottoman Syria and Palestine* (Berkeley: University of California Press, 1998).

[43] Ibrahim b. Muhammad al-Halabi (d. 1549/50), *Multaqa al-Abhur*, ed. Wahbi Sulayman Ghawji al-Albani (Beirut: Mu'assasat al-Risala, 1989), 2:336; Ibn ʿAbidin, *Radd al-Muhtar*, 4:86, 91. For more information on Shafi'i, Maliki, and Hanbali views on the validity of the writing of the mute, see ʿAbd Allah, *Ahkam al-Akhras*, 154–158.

this principle, when Hamid al-'Imadi (d. 1758), a popular Hanafi mufti in Damascus, was asked to issue a fatwa on the validity of signs in a marriage contract, his response was unambiguous:

Question: About a mute man who contracted the marriage of his daughter, who is of age, with his customary signs (*bi-ishārātihi al-ma'hūda*). The girl accepted. Is this marriage valid and can his signs replace his words?

Answer: Yes, and the issue is [discussed] in *al-Ashbah* [a legal treatise by the 16th-century Hanafi Egyptian scholar Ibn al-Nujaym].[44]

A similar logic applies to divorce. Islamic law defines marriage as a revocable contract that can be dissolved by a variety of means. The most common type of divorce is *ṭalāq*, which is the exclusive prerogative of the husband and is usually effected by his enunciation of a formulaic declaration. This custom presents an obvious problem for a man who cannot speak. The Hanafis, Shafi'is, Hanbalis, and Malikis concurred that if the mute could write, he should be encouraged to write down his wish to divorce his wife. Writing was again deemed more precise and reliable than signs, and the four schools of law reasoned that in such an important matter, ambiguity and the accompanying risk of misinterpretation should be minimized to the greatest extent possible. If a mute husband did not know how to write, then signs (*ishārāt*) constituted an acceptable substitute for spoken or written declarations. As one jurist put in no uncertain terms, "signs are the same as the words of one who can speak."[45]

Despite clear evidence of mute men's right to affirm or revoke marriage contracts, it is difficult to establish the degree to which mute people were viewed as desirable marriage partners in the eyes of the law.

44 Hamid b. 'Ali al-'Imadi (d. 1758), [*Fatawa*], as edited by Ibn 'Abidin in *al-'Uqud al-Durriyya fi Tanqih al-Fatawa al-Hamidiyya* (Cairo: al-Matba'a al-'Amira al-Miriyya, 1882/3), 1:17.

45 Al-Haskafi, *al-Durr al-Mukhtar*, 4:448. See also al-Tumurtashi, *Matn Tanwir al-Absar*, 66; Ibn 'Abidin, *Radd al-Muhtar*, 4:448. Ibn 'Abidin adds that some Hanafis and Shafi'is stipulate that a mute who can write *must* put his declaration in writing. Accordingly, he should not be permitted to sign his wish to divorce his wife; 'Ali b. Abi Bakr al-Marghinani's (d. 1196/7) seminal Hanafi textbook *al-Hidaya: Sharh Bidayat al-Mubtadi*, ed. Muhammad Muhammad Tamir and Hafiz 'Ashur Hafiz (Cairo: Dar al-Salam, 2000), 2:537; and 'Abd Allah *Ahkam al-Akhras*, 175–179. Early modern Spain had similar rules concerning the equivalence of words and signs for the purpose of consenting to marriage. See Plann, *A Silent Minority*, 18.

Nothing in the Hanafi literature I have perused indicates that they faced legally sanctioned discrimination. Shafi'i and Hanbali jurists, however, explicitly discussed the possibility of annulling a marriage contract under certain circumstances, including a spouse's contagious diseases or physical defects, such as leprosy, impotence, and other ailments.[46] For the Shafi'is, mutes fell into this category. A woman who so wished would have legal cause to demand the annulment of her marriage to a mute man, because, the Shafi'is reasoned, the impairment might be an impediment to achieving full intimacy and understanding between spouses. The Hanbalis, on the other hand, disagreed and stated flatly that muteness was not a defect (*'ayb*) precluding the validity of a marriage contract, for the purpose of marriage is the legitimation of sexual intercourse, and the inability to speak does not impede it.[47]

Inheritance, Business Transactions, and the Right to Bear Witness

In matters of inheritance, too, mute women and men suffered no formal legal discrimination. Inheritance laws were meticulously delineated and the right of a mute person to inherit his or her legally specified share of an estate or to draw up a will was not questioned. Likewise, there was no disagreement over the mute's legal right to own, purchase, sell, or otherwise dispose of property. One problem, however, might arise. Just as with marriage contracts, verbal assent was central to the conclusion of property and commercial transactions. Since mutes could not verbally consent to the terms of a contract, the jurists delineated two by-now-familiar alternative procedures: they could write, if capable of doing so, or use signs – provided that their signs could be clearly understood.[48]

[46] Although this type of annulment, called *faskh* or *tafrīq*, was technically available to both men and women, it functioned "primarily as a remedy for women who wanted to terminate a defective marriage," as men could easily divorce their wives through other means. Tucker, *In the House of the Law*, 87.

[47] 'Abd Allah, *Ahkam al-Akhras*, 167–168.

[48] Muhammad b. Hasan al-Kawakibi (d. 1685), *al-Fawa'id al-Samiyya fi Sharh al-Fara'id al-Saniyya* (Bulaq: al-Matba'a al-Kubra al-'Amiriyya, 1904/05), 2:486–489; 'Ala' al-Din Muhammad al-Haskafi (d. 1677), *al-Durr al-Muntaqa fi Sharh al-Multaqa*, printed on the margins of *Majma' al-Anhur fi Sharh Multaqa al-Abhur*, by 'Abd al-Rahman Shaykhzadah (Beirut: Dar al-Kutub al-'Ilmiyya, 1998), 4:472–473.

The same principle applied in other spheres, including the widespread practice of designating a legal proxy. The mufti Hamid al-'Imadi answered a question about whether a deaf and mute woman could legally appoint a proxy to represent her:

Question: About a man who claims the proxy of a woman who is deaf (*ṭarshā'*) and mute (*kharsā'*). Is his proxy valid despite her being described with the abovementioned attributes?

Answer: If the signs of the abovementioned woman are clear and understood, the proxy is sound [. . .].[49]

Thus the signs of the mute were deemed a legitimate alternative to the spoken and written word in some of the most crucial facets of life. But there were areas from which mute people were excluded. The most glaring example of exclusion was the presentation of testimony in court proceedings, which played an important role in both the theory and practice of Islamic jurisprudence. Of the four schools of law, the Malikis alone ruled that the written testimony of a mute was valid and that the signs of a mute who was incapable of writing, so long as they were comprehensible, constituted admissible testimony. Finding credible analogies in rulings on marriage and divorce, the Malikis reasoned that if a mute's signs were acceptable in those proceedings, it followed logically that they should be accepted in court as well.[50] The majority of Shafi'i, Hanafi, and Hanbali scholars, on the other hand, ruled that the testimony of mutes was not permissible under any circumstances. Perhaps in an attempt to balance the needs and rights of the individual, especially if male, with those of the broader community, the jurists asserted that signs are acceptable when contracting or dissolving a marriage and when buying or selling goods and property because those are activities in which, at least in their opinion, every human being must engage at some point during his or her lifetime (some basic "human rights," in today's parlance). But testifying in court, they argued, is not an absolute necessity for a person.

49 Al-'Imadi, [*Fatawa*], 1:346.

50 Muhammad b. Ahmad al-Dasuqi (d. 1815), *Hashiyat al-Dasuqi 'ala al-Sharh al-Kabir* (Cairo: Dar Ihya' al-Kutub al-'Arabiyya, n.d.), 4:167–168. 'Abd Allah, *Ahkam al-Akhras*, 312.

Since the testimony given during court proceedings might cause serious or even irreparable damage to the reputations, the material assets, and indeed the very lives of numerous individuals in a community, the jurists reasoned that the potentially negative consequences of ambiguity or misinterpretation were too grave to permit a mute's testimony through signs.[51]

A Sign Language?

When referring to the mute's acceptable means of communication, legal sources consistently employed the noun "sign" (*ishāra*) coupled with the adjectives "customary," "clear," or "understandable." But what were these "signs"? Simple nods of the head? What signs or gestures would a mute man employ in place of the standard verbal formula (usually, "I divorce you, I divorce you, I divorce you") to express his wish to divorce his wife? Unfortunately, the jurists provided no answers to these questions, perhaps because to them they were obvious. I have not encountered any descriptions of the signs used by mute people, but only perfunctory statements like "with his customary signs (*bi-ishārātihi al-maʿhūda*), that is, accompanied by his [the mute's] vocal sounds, for that is his habit."[52] Given that most congenitally deaf people are capable of articulating sounds, this terse pronouncement certainly refers to the employment of gestures, but it does not necessarily mean the use of an actual sign language.[53]

A sign language presupposes the existence of a group or community of deaf people where that language is taught. We do know of at least one such community at the very seat of Ottoman power, the deaf-mute attendants at Topkapı Palace in Istanbul. Based on painstaking research, M. Miles has convincingly debunked the tale of the "Two

[51] Ibn ʿAbidin, *Radd al-Muhtar*, 4:488; ʿAbd Allah, *Ahkam al-Akhras*, 312–315.

[52] Ibn ʿAbidin, *Radd al-Muhtar*, 4:448. See also al-Haskafi, *al-Durr al-Muntaqa*, 4:472.

[53] Sign languages are different from hand gestures in that they "have the same level of creativity as spoken languages and are in no way limited in what can be expressed. Unlike gesturing, sign languages have grammar, that is, a rich and complicated linguistic organization at the sub-lexical level, the morphological level, and the syntactic level." Ulrike Zeshan, "Sign Language in Turkey: The Story of a Hidden Language," *Turkic Languages* 6, no. 2 (2002): 232.

Brothers." According to this story, two mute brothers brought sign language to the court of Suleyman I in the 1520s. The sultan was impressed with this mode of communication, which he found dignified and solemn, and ordered that his attendants use it in the Privy Chambers. A more unlikely version is that two brothers were actually hearing and had designed a system of signs for the sole purpose of communicating privately with one other. The problem with both accounts, Miles points out, is that deaf-mute attendants served at court from at least the 1470s, and it is thus highly improbable that a sign language was introduced as a novelty in the 1520s.[54]

European authors called this sign language *ixarette*, a corruption of the Turkish word for "sign," *işaret*. It was taught at the Palace and was sufficiently sophisticated to convey complex thoughts and emotions. The Venetian diplomat Ottaviano Bon, among many others, reported: "In the *Seraglio* both the *Grand Seignor*, and divers that are about him, can reason and discourse with the *Mutes* of any thing, as well and as distinctly, *alla Mutescha*, by nods and signs, as they can with words; [...] Nay, the *Sultanas* also, and many other of the King's women do practise it, and have many dumb women and girls about them for that purpose."[55] If indeed a sign language started to be spoken at the Ottoman court in the late fifteenth century, it would be the world's oldest continuously spoken language that linguists have been able to document to date.[56]

There might have been other meeting points for deaf-mute people, possibly including a bathhouse in the Kulaksiz neighborhood of Istanbul staffed by deaf-mute attendants. The name of this neighborhood, which still exists, literally means "without ears."[57] There might also have been sign languages in the Arab provinces of the empire. Since linguists have only recently begun to research the histories of the various Arabic sign languages (e.g., Egyptian, Lebanese, Palestinian,

[54] Miles, "Deaf People," 2, 30.

[55] Ottaviano Bon, *The Sultan's Seraglio: An Intimate Portrait of Life at the Ottoman Court*, introduced and annotated by Godfrey Goodwin (London: Saqi Books, 1996), 79. Italics in the original. See also Dikici, "Imperfect Bodies," 54–55, 63–65.

[56] Ulrike Zeshan, "Sign Languages of the World," in *Encyclopedia of Language and Linguistics*, ed. Keith Brown et al. (Boston: Elsevier, 2006. 2nd ed.), 11: 358, and Zeshan, "Sign Language in Turkey," especially 238–243.

[57] Miles, "Deaf People," 55–58.

Jordanian, Syrian), their subdialects, and the extent to which they are related to one another, it is still impossible to state with certainty if any of these languages date from before the twentieth century.[58] Nonetheless, there are clues that point in that direction. First, the relatively steep rates of hereditary deafness in the Arab world that are caused by the higher prevalence of consanguineous marriages correspond to higher rates of survival of sign languages within families and communities, as they are transmitted from generation to generation. This means that "most likely there has been a long history of sign languages in the area."[59]

Second, deaf-mutes in the Arab lands could have met and hence conceivably created and taught each other sign languages. The Ottoman sultan granted high-ranking officials, such as provincial governors, special permission to hire deaf-mute servants in their local courts. We know that at least one governor in seventeenth-century Palestine availed himself of this privilege.[60] It is also reasonable to expect that after retiring from service some Palace mutes would have chosen to return to their communities of origin, including those in the Arab provinces. Furthermore, on occasion the sultan entrusted his loyal deaf-mute attendants with important missions abroad, where they may well have interacted with other mutes. In 1750, for example, "the tongueless Hasan" served as an intermediary between Istanbul and Cairo. Official documents describe Hasan as a perspicacious "special royal servant" and "privileged officeholder" who was granted the honorific

[58] Egyptian Sign Language appears to be especially well developed and widely used by the large Egyptian deaf community, estimated at two million out of a population of approximately seventy million. See M. A. Abdel-Fattah, "Arabic Sign Language: A Perspective," *Journal of Deaf Studies and Deaf Education* 10, no. 2 (2005): 212–221; Kinda Al-Fityani, "Arab Sign Languages: A Lexical Comparison," *Center for Research in Language Technical Reports* 19, no. 1 (2007): 3–13; Kinda Al-Fityani and Carol Padden, "A Lexical Comparison of Sign Languages in the Arab World," in *Sign Languages: Spinning and Unraveling the Past, Present and Future. TISLR9, Forty Five Papers and Three Posters from the 9th Theoretical Issues in Sign Language Research Conference, Florianopolis, Brazil, December 2006*, ed. R. M. de Quadros, 2–14 (Petrópolis, Brazil: Editora Arara Azul, 2008); Bernadet Hendriks, *Jordanian Sign Language: Aspects of Grammar from a Cross-Linguistic Perspective* (Utrecht, The Netherlands: LOT, 2008).

[59] Al-Fityani and Padden, "A Lexical Comparison," 4, 12.

[60] Miles, "Deaf People," 62–63 and 73–74. This privilege was apparently granted exclusively to high-ranking officials.

title of "the exemplar of those who are masters of gestures and signs" (*al-īmā' wa-l-ishārāt*).[61]

Third, a recent lexical comparison of several modern Arabic sign languages and Modern Turkish Sign Language (Türk İşaret Dili or TİD) reveals that although the word similarity index between the Turkish and Arab lists is relatively low (16–25 percent), some signs are intriguingly similar. For instance, the sign for "year" is the same in Turkish and Arabic and "is made by tapping the teeth with the index finger [. . .] because the word 'year' in Arabic is orthographically identical to the word 'tooth' [. . .]. Similarly, in TİD the sign YEAR is made by tapping the mouth with the index finger, even though in spoken Turkish there is no similar relationship between 'tooth' and 'year.'"[62] These peculiar similarities suggest that one or more Arabic sign languages might have influenced TİD since at least the Ottoman period.

Mute but Not Dumb

Islamic law made considerable efforts to guarantee and protect the rights of deaf and deaf-mute people in most spheres of life, including participation in religious rituals, the ability to marry and divorce, and the right to inherit, own, buy, and sell property. The jurists' reasoning was predicated on the assumption that the deaf and the deaf-mutes, like the hearing, did not necessarily have diminished intellectual capacities. Not only were deaf and deaf-mute people not believed to be spiritually deficient, but they were also not presumed to lack intelligence. Just how remarkable this seemingly unremarkable fact is becomes clearer when it is placed in comparative perspective. In ancient Greece, even "partial loss of hearing, because of the difficulty in verbal communication it brought on, implied partial loss of wit."[63] But far more

61 *Mühimme-i Mısır*, vol. 6, no. 601 (late Jumada I, 1163/April or May 1750). Başbakanlık Osmanlı Arşivi, Istanbul. I am grateful to Alan Mikhail for this reference.

62 Hendriks, *Jordanian Sign Language*, 35. For the sake of precision, "year" and "tooth" in Arabic are orthographically *almost* identical.

63 Martha Rose, *The Staff of Oedipus: Transforming Disability in Ancient Greece* (Ann Arbor: University of Michigan Press, 2003), 72. Similar attitudes toward deafness are also present in some of the classic texts of Judaism. See Bonnie L. Gracer, "What the Rabbis Heard: Deafness in the Mishnah," *Disability Studies Quarterly* 23, no. 2 (Spring 2003): 192–205, and Judith Z. Abrams, *Judaism and Disability: Portrayals*

damaging to the development of intelligence, it was believed, was prelingual deafness. The intellectual genesis of this attitude has often been located in the works of Aristotle. The philosopher's contention that congenital deafness constitutes a debilitating intellectual defect was rooted in a conception of the senses as functional prerequisites for the acquisition, development, and application of reason. Consequently, congenital deafness became conflated with stupidity, which over time was reflected in the vocabulary of several European languages, including English: "daft," for "stupid," derives from the Middle English "daffe" or "daft" for "deaf," and "dumb" for "mute" is still widely used today to signify "stupid."[64]

To be sure, especially starting in the sixteenth century, some European physicians, jurists, and scholars objected, arguing that deaf-mutes were no less intelligent than the hearing, but overall the legal consequences of this conflation were disastrous.[65] In early modern Spain, for example, deaf-mutes, unlike those who had lost their hearing but could speak, "were routinely classified with minors, the mentally defective, and the insane" and were denied the right to inherit their fathers' estates or to make a will.[66] And in England, well into the nineteenth century those born deaf and mute were usually regarded as either idiots or insane in criminal law proceedings, on the grounds that their defective intellect made them incapable of understanding the functioning of the legal system.[67]

In the Ottoman Arab world, too, there must have been people who believed that deaf-mutes were mentally deficient, as evidenced implicitly by some of the questions they posed to muftis concerning their legal competence. However, overall the legal, medical, historical, and biographical literature provides very little indication that loss of hearing was associated with loss of wit or that muteness, be it congenital

in Ancient Texts from the Tanach Through the Bavli (Washington, DC: Gallaudet University Press, 1998).

64 Irina Metzler, "Perceptions of Deafness in the Central Middle Ages," in *Homo debilis: Behinderte – Kranke – Versehrte in der Gesellschaft des Mittelalters*, ed. Cordula Nolte (Korb, Germany: Didymos-Verlag, 2009), 94–95.

65 Metzler, "Perceptions of Deafness," 85, 87.

66 See Plann, *A Silent Minority*, 18. For a similar assessment of the Code of Justinian and of Canon Law, see Metzler, "Perceptions of Deafness," 91–93.

67 Nigel Walker, *Crime and Insanity in England* (Edinburgh: Edinburgh University Press, 1968), 1:37, 1:224–225, 1:237–238. For early modern England, see Cockayne, "Experiences of the Deaf."

or acquired, was equated with "dumbness." Literary sources serve as corroborating evidence. The Egyptian belles-lettrist Ahmad al-Hifnawi al-Bishari (fl. 1769), for example, declared the following in a chapter of his book dedicated to "people with defects" (*aṣḥāb al-ʿāhāt*): "The blind have memory [i.e., good memorization skills], the mute intelligence, and the deaf generosity."[68]

The association of deafness and muteness with qualities like generosity and intelligence was extraordinary by contemporaneous standards. It is then perhaps not surprising that mutes could, at least in theory, aspire to one of the highest offices in the Islamic legal system, that of mufti. Muhammad Khalil al-Muradi (d. 1791) prefaced his biographical dictionary of the Hanafi muftis of Damascus from the Ottoman conquest of Syria to his own time with a discussion of the eligibility criteria for that office: an aspiring mufti must be an adult Muslim of sound mind, good character, and trusted judgment. If possessing those qualities in sufficient measure, he added, a mufti could be "a free man, a slave, a woman, a blind person, and a mute if he can write or if his signs can be understood."[69] Al-Muradi's views were based on sound precedent. Other eminent jurists before him had expressed the same opinion, concluding that "a mute can issue fatwas if his signs are clear and if he is from 'the people of the fatwa'" – in other words, if he is otherwise qualified and has received the appropriate training for the position.[70] Although I have yet to identify a single mufti who was mute (or a slave, or a woman, for that matter), it is nonetheless significant that the theoretical possibility of such employment existed.

Conclusion

Arabic biographical literature indicates that postlingual deafness, or at least societal response to postlingual deafness, was often a source

68 Al-Bishari, *Bughyat al-Jalis*, 172a.

69 Muhammad Khalil al-Muradi (d. 1791), *ʿArf al-Basham fi-man Waliya Fatwa Dimashq al-Sham*, ed. Muhammad Mutiʿ al-Hafiz and Riyad ʿAbd al-Hamid Murad, 2nd ed. (Damascus; Beirut: Dar Ibn Kathir, 1988), 13.

70 Al-Haskafi, *al-Durr al-Mukhtar*, 4:302. See also the handbook for muftis by the Syrian Shafiʿi jurist Yahya b. Sharaf al-Nawawi (d. 1277), *Adab al-Fatwa wa-l-Mufti wa-l-Mustafti*, ed. Bassam ʿAbd al-Wahhab Jabi (Damascus: Dar al-Fikr, 1988), 19.

of personal distress. A causal relationship was rarely posited between the impairment and sin or divine displeasure. Instead, its etiology was frequently ascribed to a humoral imbalance. Deafness was not an insurmountable obstacle to the exercise of personal rights, to the fulfillment of legal responsibilities, or to most other forms of participation in the social and economic life of a given community. In this respect, postlingual deafness may be best understood as a *situational* disability, shaped by variables such as age, gender, class, and other personal circumstances.

In the Ottoman Arab world, where oral communication and rhetorical sophistication were highly prized, people who were congenitally deaf certainly fared far worse and were considerably more disadvantaged in most spheres of life than those who were postlingually deaf. However, Islamic legal scholars elaborated a number of provisions that sought to establish and preserve for the *akhras* (mute) most of the fundamental rights afforded to other Muslims. These included the right to participate in religious rituals, the right to marry and divorce, the right to inherit property, and the right to engage in business and property transactions. In other words, the law, informed by an understanding of deaf-muteness that rejected associations with stupidity, strove to accommodate deaf-mute people while maintaining an equilibrium between their rights and those of the broader community.

Did Muslim jurists attempt to make similar accommodations for people with the other main sensory impairment, blindness? Did blind people's ability to hear and speak give them a comparative advantage over deaf and deaf and mute people in the hierarchy of impairments? It is to these and other questions that we turn in the next chapter.

2

Blindness

The Noblest Impairment

> Note, the last Sunday of the month of August there took place an amusement at the residence called d'Arminac in the Rue Saint Honoré, in which four blind people, all armed, each with a stick, were put in a park, and in that location there was a strong pig that they could have if they killed it. Thus it was done, and there was a very strange battle, because they gave themselves so many great blows with those sticks that it went worse for them, because when the stronger ones believed that they hit the pig, they hit each other, and if they had really been armed, they would have killed each other. Note, the Saturday evening before the aforementioned Sunday, the said blind people were led through Paris all armed, a large banner in front, where there was a pig portrayed, and in front of them a man playing a bass drum.[1]

To the modern ear, this journal entry written by an anonymous bourgeois of Paris in 1425 is chilling. His nonchalant description of an amusement that features the treatment of blind men as circus animals reflects a conception of blind people as a legitimate object of ridicule and cruelty. Although by no means universally embraced or even typical of Parisian attitudes toward blind people, these types of amusements, which signal "the most extreme relegation of

[1] Cited in Wheatley, *Stumbling Blocks,* 1. The title of the journal is *Journal d'un bourgeois de Paris*. See also Zina Weygand, *The Blind in French Society from the Middle Ages to the Century of Louis Braille*, translated from French by Emily-Jane Cohen (Stanford, CA: Stanford University Press, 2009), 16–17.

the blind to a sphere of otherness,"[2] were not exceptional at the time in French society. They were, however, absent in the Ottoman Arab world. This is not to say that blind people were never viewed with suspicion or contempt, or subjected to derision, but I have found no evidence that carefully orchestrated acts of degradation like the one recorded by the Parisian bourgeois ever took place. This chapter explores some of the reasons that might account for this striking difference.

Blindness enjoyed the most privileged place in the hierarchy of cultural meanings ascribed to physical impairments in the Arab-Islamic world. This privileged place is reflected in the special attention given to blindness in biographical and literary works. The Mamluk official Khalil al-Safadi (d. 1363), for example, composed a biographical dictionary dedicated exclusively to prominent individuals who had this impairment. It contains more than 300 biographical sketches of men who were born or became blind (no women are included). The biographies show that blind men were engaged in a wide range of occupations, among them Qur'an reciters, philosophers, lexicographers, Hadith transmitters, poets, and rulers. Al-Safadi also authored a biographical dictionary specifically about one-eyed men of note, which consists of 111 notices (again, women were excluded) and is entitled *Knowledge of the One-Eyed*. Two centuries later, the Hanafi jurist 'Ali al-Qari (d. 1606) wrote *Consoling the Blind for the Affliction of Blindness*. As the title suggests, the work was intended to comfort those who had lost their sight. To that effect, al-Qari presented and commented on a variety of favorable Qur'anic verses and Hadith sayings on the subject, and included examples of noteworthy individuals affected by blindness, such as the prophets Jacob and Shu'ayb and several companions of the Prophet Muhammad. Al-Qari emphasized forbearance in the face of adversity.[3]

[2] Weygand, *The Blind in French Society*, 17. See also Wheatley, *Stumbling Blocks*, 90.

[3] Khalil b. Aybak al-Safadi, *Nakt al-Himyan fi Nukat al-'Umyan* [The Emptying of the Pockets for Anecdotes about Blind People], ed. Tariq Tantawi (Cairo: Maktabat al-Tala'i, 1997); al-Safadi, *al-Shu'ur bi-l-'Ur* (Knowledge of the One-Eyed), ed. 'Abd al-Razzaq Husayn (Amman: Dar 'Ammar, 1988); 'Ali al-Qari, *Tasliyat al-A'ma*. Unfortunately lost is another treatise on blindness, Muhammad Ibn Tulun's (d. 1546) *Ta'jil al-Bishara li-Man Sabara 'ala Dhahab al-Basar* [Accelerating the Good Omen for Those Who Have Patience upon Losing their Eyesight].

The special status of blindness may be ascribed, in part, to its sheer frequency. Accidents, acts of violence, and congenital forms of blindness were not uncommon, but perhaps the most prevalent cause was the many blindness-inducing eye diseases, above all trachoma. Trachoma is a contagious infection of the cornea and inner lining of the eyelids. It was rife in the region because of a combination of relatively poor sanitary conditions, crowded living quarters, and hot, dry, and dusty climates that were hospitable to the bacterium *Chlamydia trachomatis*. Trachoma remains endemic in parts of Egypt to this day.

The Ottoman bureaucrat and historian Mustafa ʿAli visited Cairo from Istanbul in 1599 and observed that "most of the people of Egypt are affected by some disease and ailing. *One rarely meets a person whose eyes are bright and round*, who is [not] himself nor his male sex organ suffering from an illness, and whose physical health is manifest."[4] Mustafa ʿAli, whose overall opinion of Egyptians was rather unfavorable, if not downright hostile, might have exaggerated, but the pervasiveness of blindness and eye disease was noted in the Arabic medical literature of the time and made a vivid impression on many other travelers as well. For example, the Scottish physician Alexander Russell (d. 1768), who resided in Aleppo for thirteen years and was a keen and generally sympathetic observer of local conditions, estimated that eye diseases affected at least one in six inhabitants of the city. He observed that they struck especially fiercely during the hot summer months.[5] Indeed, Ottoman-era sources are populated with blind individuals, suggesting that this impairment was very much a part of the social and cultural landscape of Egypt and the Levant in the early modern period, just as it had been in earlier times.[6]

[4] Andreas Tietze, *Mustafa ʿAli's Description of Cairo of 1599: Text, Transliteration, Translation, Notes* (Vienna: Verlag der Österreichischen Akademie der Wissenschaften, 1975), 42–43. Italics mine. Also cited in Richardson, "Blighted Bodies," 66–67.

[5] Alexander Russell, *The Natural History of Aleppo. Containing a Description of the City, and its Principal Natural Productions in its Neighborhood. Together with an Account of the Climate, Inhabitants, and Diseases; Particularly on the Plague*, 2nd ed., revised, enlarged, and illustrated with notes by Patrick Russell (Hants., Eng.: Gregg International, 1969. Reprint of: London: Printed for G. G. and J. Robinson, 1794), 2:229, 2:298. See also al-Antaki, *al-Nuzha al-Mubhija*, 2:134–135.

[6] Blindness is the physical impairment that figures most prominently in Arabic sources from earlier times, too. As Fedwa Malti-Douglas has remarked, "The question of blindness is an important one in Islamic civilization, and appears in virtually all of

Various explanations were proffered for the causes of blindness. Some writers cited a misogynistic hadith of dubious authenticity: "If one of you has sexual intercourse, do not look at the vagina, for it causes blindness, and do not talk much, for it causes muteness."[7] In Egypt, it was believed that conceiving during the *khamsīn* season (literally "the fifty," the hot and dry winds laden with sand and dust that blow in the spring and summer for up to fifty days) could result in "deformed, blind, stunted, and lame" children, on account of the poor quality of the semen produced while the body is enfeebled by the winds.[8] Some astrologers, for their part, maintained that any child delivered during a solar or lunar eclipse would be blind at birth.[9]

Many cases of blindness resulted from ocular trauma. Some are recorded in fatwa collections because the loss of one or both eyes (or of other parts of the human body) through injury by someone else, whether intentional or unintentional, was subject to complex rules of financial compensation for bodily harm. Thus we read, for instance, of a man who hit another man with a stone and gouged out his eye, and of a small child who accidentally injured another child's eye with a stick with which he was playing.[10]

Medical compendia usually contained comprehensive discussions of ocular anatomy and ailments, in keeping with a broad historical tendency in Islamic medical literature that privileged vision as "the most noble of the external senses."[11] Physicians regarded the corruption

the major types of sources in the medieval period: from the theological and the legal through the historical to the literary and philological." Malti-Douglas, "*Mentalités* and Marginality," 215.

7 Al-Qari, *Tasliyat al-A'ma*, 31. This particular hadith was reported by Abu Hurayra. It is not included in the main Hadith collections.

8 Gary Leiser and Michael Dols, "Evliyā Chelebi's Description of Medicine in Seventeenth-Century Egypt. Part I: Introduction," *Sudhoffs Archiv* 71, no. 2 (1987): 203–204.

9 Al-Bishari, *Bughyat al-Jalis*, 171b. The astrologers' claim is also reported in al-Safadi, *Nakt al-Himyan*, 63.

10 Ibn al-Ha'ik (d. 1701/2), *Shifa' al-'Alil*, 179b, and 'Abd Allah al-Khalili al-Tarabulusi (fl. 18th c), *Fatawa*, MS, Princeton University Library, Yahuda 507, 57b, respectively.

11 Ibn Sallum, *Ghayat al-Itqan*, 42a. See also Ibn Sallum, *al-Tibb al-Jadid al-Kimiya'i*, ed. Kamal Chehadeh (Aleppo: Manshurat Jami'at Halab, Ma'had al-Turath al-'Ilmi al-'Arabi, 1997), 244; al-Antaki, *Tadhkirat Uli al-Albab*, 3:79–81; al-Antaki, *al-Nuzha al-Mubhija*, 1:306–309, 2:132–163, 3:2–16; al-Qalyubi (d. 1659), *al-Tadhkira fi al-Tibb*, 29–42; Sadaqa 'Afif (fl. 1729), *Jami' al-Ghara'ib wa-Diwan*

of the humors as a common cause of eye diseases and blindness. Human physiology was firmly grounded in the ancient Greek theory of the humors, at the core of which was a division of the cosmos into the four basic elements of fire, air, water, and earth. The human body was believed to be composed of four humors corresponding to the four elements: blood, which corresponded to air; phlegm, which corresponded to water; yellow bile, which corresponded to fire; and black bile, which corresponded to earth. When these humors were corrupt or unbalanced, illness occurred.

A simplified understanding of humoral theory was shared by the general populace as well. For example, Ahmad Tashkubrizadeh (d. 1561), a prominent scholar who held various judicial appointments, became blind as an adult. According to one biographer, the process that resulted in the loss of his vision began with his habitual use of *barsh*, an opium-based paste that enjoyed popularity in the Middle East in the sixteenth century. One day, while serving as a judge in the Turkish city of Bursa, Tashkubrizadeh, his thinking impaired by the drug, struck one of the Sultan's soldiers. The soldier and his comrades swore revenge and began to plot the judge's demise. After narrowly escaping death, Tashkubrizadeh swore off *barsh* and never ingested the paste again. Unfortunately, the sudden abstention from the highly addictive drug disrupted the humoral equilibrium between his body's "moist substances" and the drying properties of *barsh*. The "moist substances" that had been previously kept at bay by the drug began to descend from the judge's head to his eyes in increasing quantities until he became blind.[12]

Humoral disequilibrium was also the reputed cause of 'Abd al-Jawad al-Misri's (d. 1608) blindness. He was a schoolteacher in the

al-'Aja'ib, MS, Staatsbibliothek zu Berlin, We. 1193, 18a, 38a. For an overview of optics and ophthalmology in Arab-Islamic medicine, see Max Meyerhof, "The History of Trachoma Treatment in Antiquity and During the Arabic Middle Ages," *Bulletin of the Ophthalmological Society of Egypt* 24 (1936): 26–87; and G. A. Russell, "The Emergence of Physiological Optics," in *Encyclopedia of the History of Arabic Science*, ed. Roshdi Rashed (London; New York: Routledge, 1996), 2:672–715.

12 Al-Burini, *Tarajim al-A'yan*, 1:74. Tashkubrizadeh's (also transliterated as Tashkopruzade) most famous work is *al-Shaqa'iq al-Nu'maniyya fi 'Ulama' al-Dawla al-'Uthmaniyya*, a biographical dictionary of Turkish scholars, which he composed in Arabic. Al-Burini learned several details of Tashkubrizadeh's life from the latter's son, who was appointed as chief judge of Damascus during al-Burini's lifetime. The Arabic phrase for "moist substances" is *al-mawādd al-rutūbiyya*.

Biqa' Valley who moved to Damascus, where, late in life, he lost his sight. The loss was attributed to his habit of removing his turban, baring his head, and dousing it with large quantities of cold water during his preprayer ablutions. The constant exposure of his head and eyes to coldness and wetness disrupted the humoral balance in the teacher's head, resulting in blindness.[13]

The balance of the humors could be unsettled by emotional turmoil. When thieves gained entry to the room of Muhammad al-Hazrami (d. 1632/3) and stole his money and personal belongings, al-Hazrami, a Damascene scholar who was quite poor, was distraught. Shortly thereafter, he lost his sight.[14] Profound emotional distress also caused the blindness of al-Badr al-Safadi (d. after 1572), a Shafi'i mufti of Safad, who literally cried himself blind in the aftermath of a family tragedy: his eldest son accidentally – although some speculated that it was intentional – shot his younger, more promising brother, who was the father's acknowledged favorite.[15]

Considering the medical pluralism of the time, it is hardly surprising that there were as many cures for eye diseases and blindness as there were causes. If a humoral imbalance was at the root of the problem, physicians prescribed a variety of remedies and diets aimed at restoring the equilibrium. Some people believed in the special properties of certain springs, including a group of ten springs in today's southern Lebanon that were reputed to heal many different ailments ranging from leprosy to stomachaches to baldness. One of them, "the Special Spring," would give "bright, radiant eyes" to whomever drank from it for forty consecutive days: "By Allah's order they [the eyes] will be endowed with such brightness that he will be able to see the stars in the sky in broad daylight."[16]

[13] Al-Misri ultimately died as a result of dropsy, the excessive accumulation of water in the body's soft tissues, a fact that was unlikely to have been deemed a coincidence. Al-Ghazzi, *Lutf al-Samar*, 2:487–488.

[14] Al-Muhibbi, *Khulasat al-Athar*, 4:181.

[15] Al-Burini, *Tarajim al-A'yan*, 2:106–107. Al-Burini adds that he personally verified al-Safadi's blindness in the aftermath of the accident. For another example of a man who became blind from excessive crying, see the biography of Sa'id b. Jubayr in 'Abd al-Wahhab al-Sha'rani's *Lawahiq al-Anwar fi Tabaqat al-Akhyar* (Cairo: Maktabat al-Adab, 1993), 1:109–110.

[16] This is recounted by Evliya Çelebi in *Evliya Tshelebi's Travels in Palestine (1648–1650)*, translated from Turkish by St. H. Stephan (Jerusalem: Ariel Publishing House, 1980), 13.

The operation of divine forces, be they God himself, dreams or visions of the Prophet Muhammad, the intervention of holy men, the recitation of certain prayers, formulas, or Qur'anic verses, could also miraculously restore one's eyesight. A "strange thing," for example, happened to 'Umar al-Dafari (d. 1668/9), a prayer leader and jurist in Cairo: "He lost his eyesight for about twenty years, then God bestowed him with the return of his sight without treatment."[17] Similarly, Ibrahim al-Qusayri (d. 1527), an elementary schoolteacher in Aleppo, became blind as an adult. One night the Prophet appeared to him in a dream and placed his hand on one of his eyes. As a result of this intervention, partial sight was restored to him.[18] Divine forces also intervened in favor of Ahmad Ba'lawi (d. 1634), who lost his sight toward the end of his life and decided to travel to Medina for help. There, during a visit to the tomb of the Prophet Muhammad, he asked a holy man to intercede with God on his behalf, seeking the restoration of sight in at least one of his eyes. His wish was reportedly granted, for upon his return home he was approached by a man who, by unspecified means, "opened his eyes." From that moment until his death, Ahmad was able to see.[19]

The Nomenclature of Blindness

In contrast to most European languages, at least four main terms were employed in Arabic in reference to blindness.[20] *A'mā* was (and still is) the most common word for "blind." It was customarily used in legal and medical texts but appears less frequently in the biographical literature of the time, where names usually feature onomastic chains that specify the identity of a particular individual. Onomastic chains could

[17] Al-Muhibbi, *Khulasat al-Athar*, 3:220. See also the case of al-Yaman b. Mu'awiya al-Aswad, to whom God would restore the ability to see whenever he wanted to read the Qur'an, in al-Sha'rani's *Lawahiq al-Anwar*, 1:146.

[18] Ibn al-Hanbali, *Durr al-Habab*, 1:25–26.

[19] Al-Muhibbi, *Khulasat al-Athar*, 1:219. For other examples of holy men who were able to restore eyesight, see al-Sha'rani's *Lawahiq al-Anwar fi Tabaqat al-Akhyar*, translated by Virginia Vacca in *Vite e detti di santi musulmani* (Milano: TEA, 1988), 178.

[20] Malti-Douglas has observed that this is "unlike the Western languages, where the available vocabulary for the concept of blindness is actually quite restricted." In "*Mentalités* and Marginality," 219.

be lengthy and might list geographic origin, profession, and physical and intellectual characteristics. A person's full name would typically include the first name (*ism*), such as Muhammad, sometimes immediately preceded or followed by an honorific title, such as Najm al-Din (the Star of the Faith); a patronymic or series of patronymics indicating lineage (*nasab*), such as ibn Ahmad (son of Ahmad); and one or more adjectives indicating the place of origin, residence, tribe, religion, school of law, Sufi order, or profession (*nisba*). Many names also contained nicknames of all kinds (*alqāb*), including those referencing impairments, such as lame (*a'raj*), stone deaf (*uṭrūsh*), deaf (*aṣamm*), and one-eyed (*a'war*). An onomastic chain might thus read as follows: Najm al-Din Muhammad ibn Ahmad al-Ghazzi al-Dimashqi al-Shafi'i al-A'raj, translatable as "The Star of the Faith, Muhammad, son of Ahmad, originally from Gaza but settled in Damascus, a follower of the Shafi'i school of law, the lame."[21]

The word for blind individuals most frequently used in onomastic chains was *ḍarīr* (for example, Ahmad *al-ḍarīr*). *Ḍarīr* derives from the root *ḍa-ra-ra*, which means "to harm" or "to injure." It usually indicated the loss of vision through injury or illness, but on occasion it could also refer by extension to someone born blind, although the more common term for the congenitally blind was *akmah*.

Also frequently used during the Ottoman period was *baṣīr*, "seer, sighted," which was employed for a blind person renowned for notable accomplishments. Some biographers preferred the variation *al-baṣīr bi-qalbihi*, "the one who sees with his heart," if not with his eyes.[22] The words *ḍarīr* and *baṣīr*, and to a lesser extent *a'mā*, were often interchangeable, and different authors, or even the same biographer, referred to a particular blind individual as *baṣīr* in one instance and

[21] The chain could also include an agnomen (*kunya*) to indicate the person's eldest male child, for example, Abu Ahmad, "father of Ahmad." For a full discussion of onomastic chains, see Annemarie Schimmel, *Islamic Names* (Edinburgh: Edinburgh University Press, 1989).

[22] A case in point is the sixteenth-century Aleppine biographer Ibn al-Hanbali, who often used the phrase *al-baṣīr bi-qalbihi* to refer to blind individuals. It should be noted that according to Fedwa Malti-Douglas, in the Mamluk period *baṣīr* was not employed to mean "blind" but was instead used metaphorically for a person who excelled in a particular discipline or activity ("*Mentalités* and Marginality," 220). In the Ottoman-era biographical sources I have examined, on the other hand, this exclusively metaphorical use of *baṣīr* is rare.

ḍarīr in another.[23] The nicknames *baṣīr* or *ḍarīr* were not routinely affixed to the name of a person who became blind late in life, presumably because this person would not have been known as blind for most of his or her life and thus could not have acquired the nickname "blind." In those cases the expression *kuffa baṣaruhu*, "his sight became covered," often appears in the main body of the biographical notice.

There is no doubt that blindness was a physical trait that "marked" an individual and constituted an important component of his or her identity. This fact, however, should not be overstated or be interpreted as clear evidence of stigmatization. The specific context in which this nomenclature was elaborated must be underscored. As in other premodern societies, not everyone was easily identifiable by a family name, and children's names were selected from a relatively restricted menu of options – almost anyone living in seventeenth-century Cairo or Tripoli, for example, would likely have known dozens of Muhammads. As mentioned above, onomastic chains served the purpose of precise identification, and thus included occupation, honorific titles, places of origin or residence, affiliation with schools of jurisprudence or Sufi orders, and physical characteristics. Blindness was just one of many markers, signifying no more or less about an individual than the numerous other details provided in said chains.

Furthermore, the nicknames related to blindness do not necessarily bear pejorative connotations or value judgments about the individual in question. The adjective *ḍarīr* is merely descriptive, as it indicates the result of an event, however unfortunate and unenviable, that could befall anyone regardless of social status, intellectual abilities, character, or piety. In particular, words that refer to blindness do not convey a sense of absent or defective mental faculties. On the contrary, the recurrent metaphorical and euphemistic use of the term *baṣīr* or "sighted" in reference to blind individuals definitively severs the condition from connotations of darkness, occlusion, or befuddlement

[23] For example, Muhammad al-Iskandari al-Makki is *ḍarīr* in al-Muradi's *Silk al-Durar*, 4:144, but *baṣīr* in ibid., 1:154. Likewise, Ja'far al-Sanhuri is *baṣīr* in al-Ghazzi's *al-Kawakib al-Sa'ira*, 1:173, but *ḍarīr* in ibid., 3:13. The physician Da'ud al-Antaki is described as *ḍarīr* in al-Ghazzi's *al-Kawakib al-Sa'ira*, 3:150, but as *baṣīr* in al-Muhibbi's *Khulasat al-Athar*, 2:140.

and indicates that at least some of those impacted by blindness were thought to possess the power to "see" with their hearts and minds, if not with their eyes.

Prodigious Feats, Voices Like Angels, Unsightly Faces, and Greedy Hands

The euphemistic nicknames *al-baṣīr* and *al-baṣīr bi-qalbihi* ("the seer" and "the one who sees with his heart," respectively) raise the question of whether special powers were attributed to blind people. The association of blindness with supernatural forces, either beneficent or malevolent, was uncommon in the early modern Arab-Islamic world. At times, however, blind men were believed to possess an extraordinary acuity of senses other than vision. For example, Hammad al-Basri arrived in Damascus in 1529/30. He was known for his skills as a chess player, skills of such repute that the most celebrated players of Syria, Egypt, and the Hijaz traveled to Damascus to compete with him. Al-Basri not only defeated them all but also inspired awe with his ability to play five opponents on five different chessboards simultaneously. Some skeptics doubted his blindness, a suspicion that al-Basri would assuage by covering his eyes before each game. Understandably, such feats gained him great fame and, eventually, the honor of playing in the presence of the Sultan in Istanbul.[24]

Authors noted with wonder feats that highlighted blind men's prodigious awareness of their surroundings, possibly because they defied sighted people's low expectations of blind people. Abu Yazid al-Halabi (d. 1759/60), a schoolteacher, was much loved by the people of Aleppo for his piety. Piety radiated from his face, which was portrayed as shining like a lamp. Abu Yazid did not lose his sight until near the end of a very long life said to have spanned 105 years. A man who met him in Aleppo after he became blind recounted the encounter as follows: "I did not know Shaykh Abu Yazid. I went to the funeral of one of the mystics and someone pointed out the Shaykh to me. He had lost his sight. I rushed to kiss his hand." To his amazement, Shaykh Abu

[24] Muhammad Ibn Jum'a (fl. 1743), *al-Bashat wa-l-Qudah*, edited by Salah al-Din al-Munajjid in *Wulat Dimashq fi al-'Asr al-'Uthmani* (Damascus: n.p., 1949), 11.

Yazid responded: "You are Sayyid Muhammad, who lives in the store of Shaykh Muhammad al-Bunni."[25]

A similar tale is told of the Aleppo-based Qur'an instructor 'Umar al-Misri (d. 1724/5). One of al-Misri's former students, 'Umar b. Shahin, recounted two anecdotes from his adolescence that demonstrate the acuity of his blind teacher's mental faculties and other senses and the wonder that they inspired. One day 'Umar attended a private lesson with the Shaykh despite having failed to memorize the assigned lesson. In apparent fear of the scolding that his negligence might provoke, young 'Umar attempted to deceive his teacher, pulling out a small copy of the Qur'an and proceeding to recite from it. After listening carefully for some time the Shaykh pounced upon his student and snatched the Qur'an from his hand. He then proceeded to beat and berate the boy, telling him that he was deceiving not only his teacher but also himself. After young 'Umar's fear subsided, he asked the Shaykh how he had detected the ruse. The Shaykh explained: "I heard your voice coming from the room's ceiling, so I knew that there was something in your hand that prevented your voice from coming directly to my face."[26]

Biographers also remarked upon blind people's apparent ease of mobility. For instance, the abovementioned 'Umar was accompanying the Shaykh to the home of a friend. Along the way they encountered an open drain, at which point 'Umar warned his teacher to step over it. Sometime later, the pair had reason to follow the same path. The drain had since been covered, so the student issued no warning upon its approach. Nevertheless, al-Misri stopped at precisely the same point and stepped over the drain cover.[27]

Another remarkable sight was that of the blind Muhammad al-'Arbili (d. 1518) riding horses in and around Damascus "like the sighted, without anyone guiding him."[28] Similarly, Damascenes were reportedly struck by the mobility of the blind teacher Ahmad (d. 1590), who could navigate the markets and alleys of the city "better than those

[25] Al-Muradi, *Silk al-Durar*, 1:86.

[26] Ibid., 3:199–200.

[27] Al-Muradi takes this opportunity to relate similar incidents involving famous blind individuals. *Silk al-Durar*, 3:200.

[28] Ibn Tulun, *Mufakahat al-Khillan*, 2:88; al-Ghazzi, *al-Kawakib al-Sa'ira*, 1:90.

who see, and without a guide."[29] One of his students, Ba'th Allah al-Misri, who was also blind, was sometimes seen leading Shaykh Ahmad to the lecture hall of the Umayyad Mosque. Observers marveled at the sight of the blind leading the blind. Moreover, Ba'th Allah's condition did not impair his mobility in the broadest context, for he traveled from Egypt to Damascus, went on the pilgrimage to Mecca, and took trips to the Lebanese city of Tripoli as well as to Istanbul, where he performed Qur'anic recitations before the Sultan.[30]

Many of the individuals noted thus far were teachers, and it is important to recall a feature of the period's educational system, one that would in later centuries decline in scope and significance: the relationship between teacher and student. Those whose financial means enabled them to pursue an education beyond the initial stage of the *kuttāb* (roughly analogous to elementary school) would frequently contract the services of private tutors for more advanced study. This mode of instruction would engender a relationship that was both more broadly defined and more intimate in nature than today. As a result of this master–apprentice relationship, students often became *mulāzim*s, literally the attendants of their teachers, spending prolonged periods of time in the master's presence and serving him in a variety of capacities not currently associated with education. Among blind men, then, those who worked as teachers were perhaps uniquely privileged, since they enjoyed personal-professional relationships with adolescents and young men whose duties included "guide" as a matter of course.[31]

Some blind men fared quite well without any assistance. Especially impressive in this respect was the Egyptian Sufi scholar Muhammad al-Khalwati (d. 1793). He never married and was blind for the last two decades of his life. He chose to live alone, and his peers were astonished that in addition to carrying out daily activities such as cooking and

[29] Al-Ghazzi, *al-Kawakib al-Sa'ira*, 3:122.

[30] Al-Burini, *Tarajim al-A'yan*, 1:109; al-Ghazzi, *Lutf al-Samar*, 1:344–345; al-Muhibbi, *Khulasat al-Athar*, 1:453–454. For accounts of other blind travelers, see al-Ghazzi, *al-Kawakib al-Sa'ira*, 2:134–135; al-Muradi, *Silk al-Durar*, 3:101–103, 4:38.

[31] A student of the teacher 'Umar al-Misri (d. 1724/5) began studying with the blind man at age twelve. He spent most of his time with him and was known for leading his teacher "wherever he wanted" in the streets of Aleppo. See al-Muradi, *Silk al-Durar*, 3:199.

washing with no help he also raised farm animals and taught numerous students. Even more surprising was that he could point to each one of his many cats by name without being able to see them:

> He had neither relative nor stranger, neither maid-servant nor slave nor anyone to serve him in anything at all. . . . His door was always open and he had sheep, chicken, geese, and ducks and all of them went about freely in the courtyard while he attended to their fodder and feeding and watering all by himself. He cooked his food by himself and likewise washed his clothing. It spread among the people that the jinn served him – and it was not far from the truth, for he was one of the people of occult and secret knowledge. Many students came to him to study with him and learned from him. He was a decisive influence on everything and fine participation in the sciences and mystical learning, in the divine names, spirituality, awfaq, and summoning spirits. He had complete command of everything he was asked about. He had a large number of cats and he knew them individually by their names, pedigrees, and colors. He would say, "This one is Tuhfa bint Bustana, and this is Kammuna bint Yasmin, and this one is so-and-so, sister of so-and-so," and so on.[32]

Exceptional memory figures prominently among the remarkable abilities attributed to blind people. Lutfi b. Yunus was celebrated for his extraordinary intelligence and mnemonic powers. A rising star in the firmament of sixteenth-century Damascene scholars, Lutfi's promising career was temporarily stalled when he contracted an eye disease that caused him intense pain and severely degraded his vision. But Lutfi persisted in his quest for knowledge. Spurred by a strong desire to memorize the Qur'an, he hired a slave girl known for her splendid recitation of the sacred text, which he thereby memorized in its entirety. In addition, Lutfi hired several students to read aloud works drawn from a variety of disciplines, of which he memorized a considerable number, particularly in the field of belles-lettres. His aesthetic sensibility also found expression in the composition of poetry.[33]

32 'Abd al-Rahman al-Jabarti (d. 1825/6), *'Abd al-Rahman al-Jabarti's History of Egypt: 'Aja'ib al-Athar fi'l-Tarajim wa'l-Akhbar*, ed. Thomas Philipp and Moshe Perlmann (Stuttgart, Germany: Franz Steiner Verlag, 1994), 2:409–410.

33 The sources provide no additional information about the slave girl. Al-Burini, *Tarajim al-A'yan*, 3:147b-148a; al-Muhibbi, *Khulasat al-Athar*, 3:305–307. Later in life, Lutfi b. Yunus abandoned his scholarly and artistic interests in favor of what his

Blind men were also sometimes celebrated for the beauty of their voices. Several thus gifted were employed to recite the Qur'an and make the call to prayer. Shaykh Ahmad al-darir (d. 1590) was described as a pleasant and upright individual. His vocal skills were so prized that he served as a muezzin at the Umayyad Mosque, where he recited passages from the Qur'an after the sunset prayer. The angelic beauty of his voice inspired the elite of Damascus to attend his public recitations, which elicited hyperbolic praise, including comparisons with the voice of the Archangel Gabriel and a declaration that "whoever wants to listen to the Qur'an the way it was revealed should listen to the way Shaykh Ahmad al-darir recites it."[34]

Evidently, the blind man renowned for his extraordinary skills and his remarkable accomplishments in a variety of scholarly and artistic disciplines was a familiar figure in the Ottoman Arab world. Despite the genuine esteem in which many blind individuals were held, however, the mere fact of being impaired was not sufficient to gain admiration, let alone idealization. The blind, like the sighted, were perceived to be profoundly human. Personal features or qualities deemed unsavory, like those regarded as exemplary, were included in the biographical entries of blind persons just as they were in the entries of sighted persons, and it was perfectly acceptable to make quips at their expense. For instance, the Egyptian poet Abu al-Sama' al-basir (d. ca. 1655) was doubly unfortunate, being not only blind but also quite ugly. Some of his contemporaries pulled no punches, noting that the poet's physical appearance inspired revulsion and that it was much better to hear him than to see him.[35] Indeed, some Arabs found the faces of blind people, particularly those who had lost their eyesight as a result of illness or injury, unsightly. The association of sightless eyes with a repulsive appearance was probably the default cultural position, for the biographer of a personal acquaintance who had lost his vision

biographers describe as a life of hedonism. He reportedly squandered the fortune he had inherited from his father and died in poverty.

[34] The comparison to Gabriel appears in al-Burini, *Tarajim al-A'yan*, 1:109. An even more explicit association with the Archangel's voice at the Revelation appears in al-Ghazzi, *al-Kawakib al-Sa'ira*, 3:121. See also Ibn Ayyub, *al-Rawd al-'Atir*, 8, where al-Shaykh Ahmad's recitation of the Qur'an is described as so moving that it brought the audience to tears.

[35] Al-Muhibbi, *Khulasat al-Athar*, 1:129–130.

remarked with surprise: "The beauty of his eyes remained, so that one might think they could still see."[36]

But the main negative stereotype associated with blindness was avarice. Ba'th Allah al-Misri (d. 1607) received praise for his beautiful voice and moving recitation of the Qur'an, but he was also depicted as a stubborn man with an inordinately high opinion of his skills and their value in the marketplace. One of his biographers, who knew him personally, faulted Ba'th Allah for being excessively greedy and rather uncouth, because he would ask extraordinarily high fees for private recitations and haggle vigorously with his prospective patrons when the desired fee was not immediately forthcoming. Ba'th Allah's reputation for avarice was such that it inspired a famous poet to compose a satirical couplet in which he was compared to a raven.[37]

In the face of unflattering treatment, blind men often gave as good they got. An Egyptian belles-lettrist reproduced a number of jokes and anecdotes (many from earlier authors) that feature sharp rejoinders from their blind characters, as in the following story narrated by an inhabitant of Basra:

> One day I left Basra and saw a village. I spent the night there. During the night I went out to relieve myself, and there was a blind man with a jar on his shoulder and a lamp in his hand. He came to the river and filled the jar. I said to him: "You are blind, so what are you doing with a lamp?" He replied: "O busybody, I carry it for the blind of heart like you who fall on me and break my jar!"

In the same vein, "A blind man married a black woman. One day she said to him: 'If you saw my beauty, you would be amazed.' The husband replied: 'If that were the case, those who can see would not have left you to the likes of me!'"[38] The common feature of these

[36] Al-Ghazzi, *al-Kawakib al-Sa'ira*, 3:13.

[37] Al-Ghazzi, *Lutf al-Samar*, 1:344–345, 346; al-Muhibbi, *Khulasat al-Athar*, 1:453; Ibn Jum'a, *al-Bashat wa-l-Qudah*, 27. According to Fedwa Malti-Douglas, in Mamluk times there was another trope: blind men were considered especially virile. A "mode of compensation" was in place, whereby a man deprived of sight was endowed with heightened sexual prowess. A similar stereotype existed in medieval France, with the added association of excessive sexual desire with sin. Malti-Douglas, "*Mentalités* and Marginality," 226–227; Wheatley, *Stumbling Blocks*, 65. I have not encountered this trope in early modern Arabic sources.

[38] Al-Bishari, *Bughyat al-Jalis*, 172a. The association of black skin with unsightliness crops up repeatedly in Arab-Islamic literature. Variants of both anecdotes also appear

stories is an inversion of the power relationship between the sighted and the blind that highlights the latter's keen wit.

The Question of Sin

That blind men were depicted either as individuals endowed with remarkable abilities or as ordinary humans, but not as inherently spiritually or morally corrupt, may be attributed, in part, to the fact that divine punishment for sin or misdeed was notable for its absence as a main cause of blindness. Yet explanatory models prevalent in contemporaneous parts of Europe indicate that blindness was oftentimes perceived as one of the "wages of sin." One of the main debates currently engaging historians of disability in premodern Europe is the extent to which impairments were deemed to be God's punishment for sin. Some, including Zina Wygand, argue that "disability was often perceived as the visible mark of transgression or an invisible moral defect: 'The lame and the blind [. . .]; if nature has reduced them to this point and stigmatized them, it is because they have a sin to expiate.' This sin could be their own or even that of their parents [. . .]."[39] Others, notably Irina Metzler and David Turner, maintain that the connection between sin and impairment, although undeniably present, has been excessively emphasized.[40]

There is no consensus on this issue in European historiography, but the Arab-Islamic context is much clearer. The central tenets and overarching moral framework of the Qur'an display many similarities to those of the Bible, but the notion of impairment as a product of divine retribution does not feature prominently in Islamic theology. At times blindness was undoubtedly understood as a test of faith, and some verses of the Qur'an may be interpreted as connecting impairments with sin. This is perhaps most evident in the verses "What comes to you of good is verily from God; and what comes to you of ill is from

in al-Safadi, *Nakt al-Himyan*, 67. See also the spirited exchange in verse between a blind poet from Homs and a sighted colleague in which blindness was frankly discussed in al-Muradi, *Silk al-Durar*, 3:21–25.

39 Weygand, *The Blind in French Society*, 16. See also Moshe Barasch, *Blindness: The History of a Mental Image in Western Thought* (New York: Routledge, 2001), 9.

40 Metzler, "Disability in the Middle Ages," 49; Turner, *Disability in Eighteenth-Century England*, 3.

your own self (your actions)" and "Whatever misfortune befalls you is a consequence of your deeds; yet He forgives much."[41]

It is also true that some holy men, acting as God's intermediaries, were believed to be endowed with special powers that included the ability to inflict blindness on people who transgressed the perceived boundaries of Islamic mores. The Egyptian ʿAbd al-Wahhab al-Shaʿrani (d. 1565) reports a few such cases in his biographical dictionary of Sufis. For example, Suwayd al-Sinjari once saw a man eyeing a woman. When his exhortations to stop went unheeded, al-Sinjari called upon God to blind the man, who lost his sight on the spot. Some days later, the blinded man repented and asked for forgiveness. Al-Sinjari took pity on him and asked God to restore his sight, except when he engaged in illicit behavior. God obliged the request, and from that moment forward the man would become temporarily blind whenever he looked at anything prohibited.[42]

Overall, however, the link between impairment and sin was decidedly de-emphasized. This is first and foremost a function of the absence in Islamic theology of the doctrine of original sin, of humans as inherently sinful beings, which is central to Christianity. In the case of blindness, it is also certainly the result of how this impairment was represented in the foundational texts of Islam, the Qur'an and the Hadith.

Blindness in the Qur'an and Hadith

Blindness is by far the most frequently mentioned impairment in the Qu'ran. It appears in 48 verses out of more than 6,000, whereas

41 Ahmed Ali, *Al-Qur'an: A Contemporary Translation* (Princeton, NJ: Princeton University Press, 1988), 84 (Sura 4, v. 79), and 415 (Sura 42, v. 42), respectively. Henceforth cited as *Al-Qur'an*. For a fuller discussion of the link between impairments and sin in Islamic theology, see Ghaly, *Islam and Disability*, 42–46, and Dols, *Majnūn*, 246.

42 Al-Shaʿrani, *Lawahiq al-Anwar*, translated by Virginia Vacca in *Vite e detti di santi musulmani*, 195. For more examples, see ibid., 37, where a holy man asks God to blind a woman who is lying, and pp. 247 and 341. Al-Muhibbi's biographical dictionary includes a brief entry on Yusuf al-Ayyubi (d. 1656/7), a wealthy employee of the chief judge's court in Damascus. Al-Ayyubi became blind near the end of his life, and some people attributed the onset of his condition to a vile oath he had uttered during a quarrel some years previously (*Khulasat al-Athar*, 4:508).

deafness and muteness are cited seven times, and lameness twice. The verb *'amiya*, "to be or become blind," and its derivates are employed to denote the state of being or the process of becoming blind in two distinct contexts: literal, to indicate the physical absence or loss of vision, and metaphorical, to refer to the absence or loss of spirituality. Of the two, the metaphorical meaning is most frequent.[43]

When mentioned in conjunction with deafness and muteness, blindness signifies deficiency in faith and the incapacity or unwillingness to understand, to "see" the truth of revealed religion. For example, in Sura 41 ("Adoration"), the tribe of Thamud that lived in the northern Hijaz and rejected the message of the Prophet Muhammad is described in the following way: "As for the (tribe of) Thamud, We tried to guide them, but they preferred blindness to guidance; then they were seized by the torment of a humiliating punishment as requital for their misdeeds.[44] Similarly, in Sura 2 ("The Cow") the "hypocrites" who merely pretend to believe in God are thus depicted: "They are like a man who kindles a fire, and when its glow has illuminated the air God takes away their light leaving them in the dark where they will not be able to see. They are deaf, dumb, and blind, and shall never return."[45]

Over time, the metaphorical interpretation of blindness as deficient capacity for belief remained present in certain literary genres, particularly in dream manuals. A classic early modern example is *Ta'tir al-Anam fi Ta'bir al-Manam* (The Perfuming of Humankind through the Interpretation of Dreams) by the renowned scholar 'Abd al-Ghani al-Nabulusi (d. 1731). Among thousands of entries are most parts of the human body, accompanied by their symbolic meaning when featured in dreams. The eye is the subject of an especially lengthy entry.

43 Bazna and Hatab, "Disability in the Qur'an," 14.

44 *Al-Qur'an*, 408 (Sura 41, v. 17).

45 *Al-Qur'an*, 13 (Sura 2, vv. 17–18). See also Sura 2, v. 171: "The semblance of the infidels is that of a man who shouts to one that cannot hear more than a call and a cry. They are deaf, dumb, and blind, and they fail to understand;" Sura 5, v. 71: "And [the children of Israel] imagined that no trials would befall them; and they turned deaf and blind (to the truth). But God still turned to them; yet many of them turned blind and deaf again; but God sees everything they do;" and Sura 10, vv. 42–43, again concerning those who refuse God's word: "Some of them listen to you: But can you make the deaf hear who do not understand a thing? Some of them look toward you: But can you show the blind the way even when they cannot see?" See also Andrew Rippin, "Seeing and Hearing," in *Encyclopaedia of the Qur'an*, ed. Jane McAuliffe (Leiden: Brill, 2004), 4:573–576.

Echoing a long and well-established tradition of Muslim dream manuals, al-Nabulusi begins by explaining that the eye signifies primarily the dreamer's religiosity, hence the appearance of numerous eyes on one's body indicates "an increase in religion."[46] A separate entry dedicated to blindness states that it may symbolize either a "straying in religion" or the accumulation of wealth. Thus, al-Nabulusi argues, a man who sees himself as blind in a dream will either "forget the Qur'an" or become wealthy.[47] At the purely metaphorical level, then, visual impairment signifies an individual's privileging of the profane over the sacred.

On the other hand, the rare occasions when the word *a'mā* (blind) is used in the Qur'an to describe individuals who lack the physical ability to see evince an attitude that is explicitly inclusive and accommodating. A crucial verse that is habitually cited to this day in discussions about disability appears in Sura 24 ("The Light"). God calls upon the community of the faithful to refrain from excluding the blind, the lame, and the sick from sharing their meals:

> There is no harm if the blind, the lame, the sick, or you yourselves, eat in your own houses or the houses of your fathers, mothers, or your brothers' houses, or those of your sisters, or your fathers' brothers' or sisters', or your mothers' brothers' or sisters', or in the houses whose care is entrusted to you, or the houses of your friends.[48]

This verse suggests that in seventh-century Arabia, when the Qur'an was revealed, there must have been some discrimination against people who were blind, sick, or had mobility impairments. The reason sharing meals with them should be avoided, however, is not explained, and it was not clear to Qur'anic exegetes. In the opinion of a well-known thirteenth-century commentator, in pre-Islamic times some people were repulsed by "the wandering hand of the blind" (presumably a reference to touching the food they could not see), by "the smell

46 Al-Nabulusi, *Ta'tir al-Anam*, 2:82.

47 Ibid., 2:85. The association between blindness and either straying in religion or becoming wealthy, two entirely different significations, one negative and the other positive, is somewhat puzzling. We can speculate that there might be a compensation system at work: you lose your eyes but you gain material wealth.

48 *Al-Qur'an*, 304 (Sura 24, v. 61). This is in a way reminiscent of Jesus, who, the Gospel of Luke reports, exhorted his listeners to invite the poor, the crippled, the lame, and the blind to their banquets.

and illness of the sick," and they did not enjoy sitting with the lame. The verse, he explained, was intended to put an end to a practice that was caused by the haughty mores of pre–Islamic Arabs.[49]

A more magnanimous explanation proposed by another Qur'anic scholar was that pre–Islamic Arabs "used to feel too embarrassed to eat with the blind, because they could not see the food or where the best morsels were, so others would be able to take the best pieces before they could. They felt too embarrassed to eat with the lame because they could not sit comfortably, and their companions might take advantage of them, and they felt embarrassed to eat with the sick because they might not eat as much as others. So they were afraid to eat with them lest they were unfair to them in some way."[50] Yet another interpretation proposed that the verse refers to the blind, lame, and sick's "refraining from eating with the healthy for fear that they [the healthy] would find them filthy."[51] It is noteworthy that Qur'anic exegetes could not reach a consensus on the correct explanation of this verse, for it suggests that cultural attitudes toward blindness had changed to the point that medieval commentators found it difficult to relate to the distaste that their pre–Islamic and early Islamic forebears felt for people who were blind or had other physical impairments.

But the most important and unambiguous statement concerning blindness is in Sura 80 ("He Frowned"), which narrates an incident involving the Prophet Muhammad. One day, while in Mecca, Muhammad was preaching the message of Islam to a group of pagan notables of the Quraysh clan whom he was trying to convert. The Prophet's address was interrupted by a man named 'Abd Allah Ibn Umm Maktum, a cousin of Muhammad's first wife Khadija. 'Abd Allah Ibn Umm Maktum, who was born blind, was one of the first converts to Islam and a faithful follower of the Prophet. On the day in question he walked into the gathering, causing some disruption. Muhammad

[49] Muhammad al-Qurtubi (d. 1273), *al-Jami' li-Ahkam al-Qur'an* (Cairo: Dar al-Kitab al-'Arabi li-l-Tiba'a wa-l-Nashr, 1967), 6:313.

[50] Isma'il Ibn Kathir (d. 1373), *Tafsir Ibn Kathir (Abridged)*, abridged by a group of scholars under the supervision of Safi-ur-Rahman al-Mubarakpuri (Riyadh: Darussalam, 2000), 7:125–126.

[51] 'Abd Allah al-Baydawi (d. ca. 1386), *Anwar al-Tanzil wa-Asrar al-Ta'wil* (The Lights of Revelation and the Secrets of Interpretation) (Beirut: Dar al-Kutub al-'Ilmiyya, 1999), 2:131.

frowned and turned away from ʿAbd Allah, displaying his displeasure at the interruption and betraying impatience with the blind man. Soon thereafter, God revealed the following verses to Muhammad:

> He frowned and turned away, because a blind man came to him. What made you think that he will not grow in virtue, or be admonished, and the admonition profit him? As for him who is not in want of any thing, you pay full attention, though it is not your concern if he should not grow (in fullness). As for him who comes to you striving (after goodness), and is also fearful (of God), you neglect.[52]

The meaning of God's reprimand is that the physical state of blindness does not affect one's status as a Muslim, it implies no deficiency in intellectual capacity, moral rectitude, or spiritual awareness, and it does not constitute an impediment to one's full participation in the community of believers.

We know through the Hadith that following this incident ʿAbd Allah Ibn Umm Maktum became a trusted and cherished companion of the Prophet Muhammad, who was so mortified by his own reproachable behavior that he wished those verses could be expunged from the Qur'an.[53] Muhammad appointed him to serve in the prestigious roles of muezzin and prayer leader to the community of the faithful in Medina during his absences from the town. Moreover, despite a verse in Sura 48 ("Victory") that exempts the blind, the lame, and the sick from the otherwise mandatory participation in warfare for the cause of God (*jihād*), so ardently did ʿAbd Allah Ibn Umm Maktum wish to fight alongside his people that he was granted the honor of serving as standard-bearer in military campaigns.[54] He eventually died

52 *Al-Qur'an*, 522 (Sura 80, vv. 1–10).

53 Scott Kugle, "Vision and Blindness," in *Encyclopaedia of the Qur'an*, ed. Jane McAuliffe (Leiden: Brill, 2005), 5:447.

54 "It is not binding on the blind, the lame, or the sick (to follow this command); but those who obey God and His Apostle will be admitted by God to gardens with running streams; but those who turn back will suffer a painful doom." *Al-Qur'an*, 440 (Sura 48, v. 17). This exemption is not offered because they were perceived to be polluted or because of other characteristics that made their participation doctrinally unacceptable, but as an acknowledgment of the practical difficulties they would encounter in war. A similar message also appears in Sura 4, v. 95: "The faithful who sit idle, other than those who are disabled [*ghayr ulī al-ḍarar*], are not equal to those who fight in the way of God with their wealth and lives. God has exalted those

clutching the flag of the Muslims during a fierce battle against the Persians in 636 CE.

The Hadith contains several other sayings that confirm the privileged place of blindness in the hierarchy of impairments. The following parable was narrated by Abu Hurayra (d. 681), a companion of the Prophet Muhammad who claimed to have heard it from the Prophet himself. God decided to test three poor men: a leper, a blind man, and a bald-headed man (due to a disease). He sent an angel who asked them what they wished for the most. The leper replied: "Good color and good skin, for the people have a strong aversion to me." The bald-headed man said: "I like good hair and wish to be cured of this disease, for the people feel repulsion for me." The blind man wished that God would "restore my eye-sight to me so that I may see the people." The angel touched each one of them and their wishes were granted. He also gave each of the three men a pregnant animal, so that they would multiply and bring wealth to their owners. Some time passed.

Then the angel, disguised in the shape and appearance of a leper, went to the leper and said, "I am a poor man, who has lost all means of livelihood while on a journey. So none will satisfy my need except Allah and then you. In the Name of Him Who has given you such nice color and beautiful skin, and so much property, I ask you to give me a camel so that I may reach my destination." The man replied, "I have many obligations (so I cannot give you)." The angel said, "I think I know you; were you not a leper to whom the people had a strong aversion? Weren't you a poor man, and then Allah gave you (all this property)." He replied, "(This is all wrong), I got this property through inheritance from my fore-fathers." The angel said, "If you are telling a lie, then let Allah make you as you were before."

Then the angel, disguised in the shape and appearance of a bald man, went to the bald man and said to him the same as he told the first one, and he too answered the same as the first one did. The angel said, "If you are telling a lie, then let Allah make you as you were before."

The angel, disguised in the shape of a blind man, went to the blind man and said, "I am a poor man and a traveler, whose means of livelihood have been exhausted while on a journey. I have nobody to help me except Allah, and after Him, you yourself. I ask you in the Name of Him Who has given you

in rank who fight for the faith with their wealth and souls over those who sit idle [...]."

back your eye-sight to give me a sheep, so that with its help, I may complete my journey." The man said, "No doubt, I was blind and Allah gave me back my eye-sight; I was poor and Allah made me rich; so take anything you wish from my property. By Allah, I will not stop you for taking anything (you need) of my property which you may take for Allah's sake." The angel replied, "Keep your property with you. You (i.e., 3 men) have been tested and Allah is pleased with you and is angry with your two companions."[55]

This parable shows in unambiguous terms that the blind is more honest, generous, and compassionate than the leper and the bald-headed man. Thus God rewards his moral superiority.

In sum, the Qur'anic verses concerning blindness, the examples of 'Abd Allah Ibn Umm Maktum and that of 'Itban b. Malik (another blind companion of the Prophet depicted positively in the Hadith), and the parable above could not have failed to inform, at least to some extent, perceptions of and attitudes toward blind people as the Muslim community expanded.

In the Eyes of the Law

Considering that the Qur'an and the Hadith were the two most important textual sources of Islamic legal thought, it is not surprising that Islamic jurisprudence elaborated laws concerning blindness that encouraged blind people's full participation in the life of their communities. Some jurists even dedicated chapters of their law books to rulings pertaining specifically to the visually impaired.[56] Islamic law made a distinction between those who were completely blind and those who were only blind in one eye. People who were partially blind were simply not considered blind as far as the law was concerned, while who were completely blind had the same rights and obligations as those who could see in most, but not all, spheres of life.

[55] Muhammad b. Isma'il al-Bukhari (d. 870), *[Sahih al-Bukhari]: The Translation of the Meanings of Sahih Al-Bukhari: Arabic-English*, translated by Muhammad Muhsin Khan. 2nd rev. ed. (Medina, Saudi Arabia: Islamic University, 1973–1976), Volume 4, Book 56, Number 670.

[56] See, for example, the Egyptian Hanafi jurist Zayn al-'Abidin Ibn Nujaym's (d. 1563) terse chapter on the blind in his manual entitled *al-Ashbah wa-l-Naza'ir*. Edited with a commentary by Ahmad b. Muhammad al-Hamawi (d. 1686/7) in *Ghamz 'Uyun al-Basa'ir Sharh Kitab al-Ashbah wa-l-Naza'ir* (Beirut: Dar al-Kutub al-'Ilmiyya, 1985), 3:344–345.

Blind men and women married, had children, divorced, and were involved in custody and a host of other disputes just like people who could see. On occasion, a woman married to a blind man sought to disentangle herself from an unhappy marriage by trying to use the excuse of her husband's blindness (a man could divorce a woman anytime for any or no reason and thus would not need to find pretexts). For example, Hasan al-Husayni (d. 1811), the chief Hanafi mufti of Jerusalem, was once asked to issue a fatwa in response to this question:

Question: About a man who married a woman and lived with her in a village near the city, then for a while lived with her in the city, and [now] wants to return to the village. She refuses to move [back to the village] with him. He paid her the prompt dower and he is legally responsible for her. Is she obligated to move to the village with her husband, *even if he is blind*, or not?

Answer: Yes, the aforementioned wife is obligated to move to the village with her aforementioned husband, since he is legally responsible for her, *even if he is blind, because he is like the sighted except in certain matters, and this is not one of them*, as is reported in the books of the [Hanafi] school. This is the situation, and God knows best.[57]

Evidently the wife did not want to leave the city and hoped that her husband's blindness would constitute a legal loophole to the rule that the wife must follow her husband. She was unsuccessful because according to Hanafi law so long as the husband fulfilled the legal obligations stipulated in the marriage contract by paying the required dower, it was of no consequence whether he was sighted or not.

It was equally irrelevant if the blind spouse was the wife, as is evident in another question posed to the same mufti of Jerusalem. The husband of a blind woman had married a second wife. The first wife had left her husband after quarrelling with him and had returned to the house of her family, who was now refusing to send her back to her husband. Was her family allowed to do that? The mufti answered unequivocally that so long as the husband could provide his wife with a legally suitable dwelling (in compliance with the rules of cohabitation of co-wives, which stipulated that they be given separate living quarters),

57 Al-Husayni (d. 1811), *Fatawa*, 20a. Emphasis mine.

her family was obligated to return her to him, even if the authorities had to resort to force to compel them to do so. It made no difference if she was blind or sighted, added the mufti.[58]

There were, however, some areas in which the blind and the sighted did not have the same rights and obligations. Blind people had the duty to pray, like everyone else, but they were not required to attend the most important weekly communal prayer, the Friday prayer. Hanafi law stipulated eleven categories of persons excused from attendance: (1) Those in a location from which the call to prayer could not be heard; (2) those in poor health; (3) slaves; (4) females; (5) children; (6) the insane; (7) the blind; (8) the lame; (9) prisoners; (10) those living in fear (of rulers, thieves, and the like); and (11) those experiencing severe weather.[59] This exemption for people with physical and sensory impairments was not predicated on the notion that physical imperfection denotes pollution, corruption, or ritual impurity, and sight was not deemed a prerequisite for communication with the divine through prayer or considered an impediment to understanding the lessons imparted during the Friday sermon. Rather, its purpose was pragmatic: attendance should not be required of those for whom it might constitute an impractical, unreasonable, or impossible task. This included blind people, who might find it difficult to make their way to the congregational mosque.

Further evidence that the reasoning underpinning the exemption was pragmatic in nature is contained in the legal compendium of Ibn 'Abidin (d. 1836). The jurist argued that it applied only to those who were entirely sightless, not to those with only one eye or those with weak eyesight. Ibn 'Abidin also stated that a blind man ought in fact to attend the Friday prayer if he could procure a guide to the mosque. He immediately added that, based on his own observation of blind men in Damascus, attending the Friday communal service was mandatory "for some blind people who walk in the markets and know their way about without anyone guiding them and without discomfort, and who

[58] Ibid., 32a.

[59] Al-Tumurtashi, *Matn Tanwir al-Absar*, 30; al-Haskafi, *al-Durr al-Mukhtar*, 3:26–29; Ibn 'Abidin, *Radd al-Muhtar*, 3:26–29. The same rationale accounts for why blind people were not obligated to perform the pilgrimage to Mecca or to participate in *jihād* (war against non-Muslims).

know how to get to any mosque they wish without having to ask for directions."[60]

Under certain circumstances blind men were also permitted to exercise leadership in prayer (*imāma*). Other religious traditions were more restrictive. According to the Old Testament, no one with a "blemish" can serve as a priest. Those with such disqualifying blemishes are named therein: the blind and those with eye blemishes, the lame, dwarfs, people with flat noses, people with broken feet or hands, crooked backs, crushed testicles, and who have scurvy or scabs. The Old Testament asserted that people whose bodies had certain imperfections were in a state of ritual impurity and would defile any religious service over which they officiated.[61]

The consensus of opinion among Muslim jurists on this point represents a considerable departure from the Old Testament. This may be explained by two factors. First, in accordance with the message of the Qur'an and Hadith, the scholars of Islamic law generally rejected any link between physical characteristics and spiritual qualities. Second, in Sunni Islam there are no priests who serve as intermediaries between God and believer. An imam does not mediate the prayers of others but merely stands at the "front" of the congregation (the literal meaning of the term imam) and guides them in prayer.

60 Ibn 'Abidin, *Radd al-Muhtar*, 3:29. Other Hanafi jurists who preceded him asserted that the Friday prayer, the pilgrimage to Mecca, and participating in *jihād* were not obligatory for the blind even if he or she could procure a guide. See Ibn Nujaym, *al-Ashbah wa-l-Naza'ir*, 3:344.

61 Leviticus 21:16–23 (King James Version): "And the Lord spake unto Moses, saying / Speak unto Aaron, saying, Whosoever he be of thy seed in their generations that hath any blemish, let him not approach to offer the bread of his God. / For whatsoever man he be that hath a blemish, he shall not approach: a blind man, or a lame, or he that hath a flat nose, or any thing superfluous, / Or a man that is brokenfooted, or broken-handed, / Or crookbacked, or a dwarf, or that hath a blemish in his eye, or be scurvy, or scabbed, or hath his stones broken; / No man that hath a blemish of the seed of Aaron the priest shall come nigh to offer the offerings of the Lord made by fire: he hath a blemish; he shall not come nigh to offer the bread of his God. / He shall eat the bread of his God, both of the most holy, and of the holy. / Only he shall not go in unto the vail, nor come nigh unto the altar, because he hath a blemish; that he profane not my sanctuaries: for I the Lord do sanctify them." For a thorough analysis of the ways in which Biblical writers stigmatized people with disabilities, see Saul M. Olyan, *Disability in the Hebrew Bible: Interpreting Mental and Physical Differences* (New York: Cambridge University Press, 2008).

There were, however, some restrictions. When the Hanafi Palestinian mufti Khayr al-Din al-Ramli (d. 1671) was asked "Is the *imāma* of the blind discouraged or not if there is no one better than him to do it?" he replied that Hanafi law neither forbade nor discouraged the *imāma* of the blind unless a person deemed more knowledgeable was present. In support of this ruling he cited the precedents of ʿAbd Allah Ibn Umm Maktum and ʿItban b. Malik, both blind Companions of the Prophet who frequently served the first community of believers in this capacity. Al-Ramli also mentioned as supporting evidence the consensus of Shafiʿi jurisprudence, which made no distinction at all between the blind and the sighted in this context.[62] In fact, Shafiʿi jurists actually privileged blind men as imams because their lack of vision enabled them to devote their undivided attention to praying and made them less likely to be distracted.[63]

Al-Ramli's ruling was in full conformity with Ottoman-era Hanafi law books, according to which a blind man can indeed lead the community of the faithful in prayer, but only in the absence of a man more knowledgeable than him. A large number of criteria could be applied to establish an individual's status as "more knowledgeable," but complete mastery of the rules of prayer and excellence in reciting the Qur'an were deemed most relevant. Since many blind men, as we shall see below, mastered the Qur'an, it is not surprising that some did in fact serve as imams.[64]

The principal arena of the law from which the blind were excluded was testimony in court proceedings. Another fatwa issued by Khayr al-Din al-Ramli encapsulates the reasoning for this exclusion. The question he posed concerned the rightful ownership of a vegetable garden. The petitioner contended that the property was a *waqf* (pious endowment). He presented two witnesses in support of his claim, one of whom was blind. Although al-Ramli ultimately was not able to

[62] Al-Ramli, *al-Fatawa al-Khayriyya*, 1:10. See also a fatwa by Ibn al-Ha'ik in *Shifa' al-ʿAlil*, 2a.

[63] Ghaly, *Islam and Disability*, 108.

[64] Ibn Nujaym, *al-Ashbah wa-l-Naza'ir*, 3:344; al-Tumurtashi, *Matn Tanwir al-Absar*, 18; al-Haskafi, *al-Durr al-Mukhtar*, 2:294, 298; Ibn ʿAbidin, *Radd al-Muhtar*, 2:294, 298–299. See also Ghaly, *Islam and Disability*, 108–109. For examples of blind prayer leaders, see al-Ghazzi, *al-Kawakib al-Sa'ira*, 1:56; 1:214; 3:176; al-Muhibbi, *Khulasat al-Athar*, 3:220.

resolve the problem of the property's ownership, he did address the issue that concerns us here, the validity of a blind man's testimony in court. Responding categorically in the negative, he explained that a blind man's testimony must inevitably rely exclusively on what he hears rather than what he sees. Hence it constitutes hearsay and is inadmissible. In other words, because in court proceedings the blind can only distinguish the accuser from the accused on the basis of their voices, confusion might ensue, which might in turn result in judicial error and the ultimate miscarriage of justice, conviction of the innocent. The potential consequences of ambiguity or misinterpretation were deemed too serious to allow blind people to testify.[65]

Making a Living

There was nothing in Islamic law prohibiting blind people from earning a living, and many blind men did in fact work. One of the most common avenues of employment was that of muezzin, the man who calls Muslims to prayer from a mosque five times a day. Prayer leader, Qur'an teacher, and Qur'an reciter were other occupations frequently pursued by the blind. The latter was the most remunerative because of the high demand for professional reciters. In addition to working in mosques, professional Qur'an reciters performed at funerals, festivals, holidays, and in people's homes. Blind men were often favored over sighted men in private homes because they could not see the women of the house.

The main reason for the relatively high proportion of blind men engaged in these professions is central to understanding the privileged place of blindness within hierarchies of impairment: the paramount role of orality in the transmission of knowledge. This is true as a general observation, but it applies more particularly to the fields of Qur'anic memorization and recitation. The centrality of the Qur'an to the social and cultural life in Muslim communities need not be

[65] Al-Ramli, *al-Fatawa al-Khayriyya*, 2:23. See also 2:30–31. A similar question, which prompted the same response, was posed to the mufti Muhammad al-Tumurtashi (d. 1596), *Fatawa*, 97b-98a. See also Ibn 'Abidin, *Radd al-Muhtar*, 2:30, 8:193; Ibn Nujaym, *al-Ashbah wa-l-Naza'ir*, 3:344; al-Tumurtashi, *Matn Tanwir al-Absar*, 152; al-Haskafi, *al-Durr al-Mukhtar*, 8:193. One of the few exceptions to this general rule accepts the testimony of the blind on issues of his own lineage (*nasab*).

belabored here, but it is essential to appreciate the primacy of the oral, as opposed to the written, mode of transmission of scripture:

> Learning the Qur'an means learning the correct sound of the Qur'an. The science of *tajwid* [Qur'an cantillation] is itself transmitted orally, the student imitating and practicing the sounds produced by the teacher. Since many of the rules for pronunciation are uniquely applied to the Qur'an and are not applicable to the literary language or to the colloquial dialect, they are incomprehensible without oral example.[66]

This cultural factor is particularly significant because it explains why blind men did *not* face insurmountable obstacles in pursuing occupations in which the aural was of greater import than the visual and textual.

Blind people achieved success in various other fields of scholarship besides prayer leader, Qur'an reciter, and Qur'an teacher. Da'ud al-Antaki (d. 1599), arguably the most famous Arab physician of the Ottoman period, was blind.[67] Three of the most celebrated Arab poets of the seventeenth century, Musa al-Ram Hamdani and Qasim al-Halabi from Aleppo, and Abu al-Sama' from Egypt, were blind.[68] Numerous blind men were renowned grammarians, scholars, and jurists. The Aleppine 'Abd al-Karim al-Sharabati (d. 1764) was a prominent Shafi'i scholar who specialized in Hadith. When just thirty years old he lost his eyesight within days of his father's death. Undeterred by the double misfortune, al-Sharabati dedicated the next forty years of his life to the zealous pursuit of his scholarly interests and continued to travel abroad for advanced study with some of the most acclaimed teachers of his time. He authored many works, including his memoirs, which he dictated a year before his death. Primarily an intellectual autobiography, they reveal the breadth of the scholarly network to which he belonged.[69] Another example of a renowned blind scholar

[66] Kristina Nelson, *The Art of Reciting the Qur'an* (Austin: University of Texas Press, 1985), 15. See also William A. Graham, *Beyond the Written Word: Oral Aspects of Scripture in the History of Religion* (New York: Cambridge University Press, 1987).

[67] Al-Muhibbi, *Khulasat al-Athar*, 2:140–148.

[68] Ibid., 4:435–442, 1:129–130, 3:377. Another blind poet was 'Uthman al-Ma'arrawi al-Himsi. Some of his poetry is reported in al-Muradi, *Silk al-Durar*, 3:21–25.

[69] 'Abd al-Karim al-Sharabati, *al-Thabat al-Mubarak*, MS, Chester Beatty Library, Dublin, Ar. 4273. For a biographical sketch of al-Sharabati, see al-Muradi, *Silk al-Durar*, 3:70–72.

is Tashkubrizadeh, whose blindness was attributed to the abrupt cessation of his drug habit. After losing his sight he resigned his judicial appointment and dedicated the remainder of his long life to scholarship. Tashkubrizadeh earned fame for his prolific and varied endeavors in the fields of logic, philosophy, biography, and poetry.[70]

Perhaps an even more remarkable accomplishment for a blind man was a successful career as mufti, a prestigious post in the judicial establishment. Muhammad Ibn Sultan (d. 1544) is a case in point. He became blind during his appointment as chief Hanafi mufti of Damascus. Undeterred by his changed condition, he persevered in his legal endeavors. To prevent forgery and ensure that the wording of his rulings could not be modified, he would stamp each fatwa with an engraved seal bearing his name. This very same strategy was also employed by one of his colleagues, a blind Hanafi mufti in Jerusalem.[71]

The impairment also did not preclude the appointment to administrative positions in pious endowments (sing. *waqf*). In seventeenth-century Damascus, a blind man held the post of collector of revenues for a large *waqf*, and another worked as the comptroller of a prestigious school. The duties of the latter included purchasing and controlling the quality of the food served in the school.[72] At times, concerns about the ability of blind people to fulfill their responsibilities, or perhaps the baser desire to use the impairment as an excuse to take over their jobs, inspired challenges to their appointment. A question posed to the mufti Hamid al-ʿImadi presents the case of the superintendent of a family *waqf* who had suddenly lost his eyesight. Although there was no evidence that the impairment affected his job performance, some of the *waqf*'s beneficiaries requested the superintendent's removal solely on the basis of his blindness. The mufti flatly refused this request. Basing his decision on ample precedent, he asserted that blindness

70 Al-Burini, *Tarajim al-Aʿyan*, 1:76. For more examples of blind scholars, see Ibn Tulun, *Mufakahat al-Khillan*, 1:263–264; al-Ghazzi, *al-Kawakib al-Sa'ira*, 1:72, 2:134–135, 2:118, 3:194; al-Muhibbi, *Khulasat al-Athar*, 1:81–82, 2:357, 4:181, 4:308–309.

71 Ibn Sultan's biography is recorded in al-Ghazzi, *al-Kawakib al-Sa'ira*, 2:12–14, 3:13; al-Muradi, *ʿArf al-Basham*, 29–31. The blind Hanafi mufti of Jerusalem was ʿAbd al-ʿAziz al-Maqdisi (d. 1542). Al-Ghazzi, *al-Kawakib al-Sa'ira*, 2:170.

72 Taisir Khalil Muhammad El-Zawahreh, *Religious Endowments and Social Life in the Ottoman Province of Damascus in the Sixteenth and Seventeenth Centuries* (Karak, Jordan: Mu'tah University, 1992), 145, 146.

alone is insufficient cause for barring an individual's appointment as *waqf* administrator.[73]

Poverty, Begging, and Charity

To a greater or lesser degree, the professions described above were within the province of the educated. Because Ottoman-era biographical sources tend to provide limited information about individuals outside the world of scholarship, it is much more challenging to gauge what employment opportunities were available to uneducated blind people, to those who sought work in the crafts, trades, or farming, and above all to blind women of all levels of education. It is likely that, as in other premodern societies, those persons often did not fare well. Islamic law, however, made some provisions for them through the system of *nafaqa* (maintenance). Moreover, the father, or in the absence of the father, the closest male relative of a woman, minor child, man with a chronic illness, or blind man were enjoined to provide material support in the form of money, housing, food, or clothing to the needy relative. Thus, when a mufti in eighteenth-century Tripoli was asked for a legal opinion on whether a rich uncle should financially support the young children of his blind and destitute brother and his sighted but equally destitute wife, the answer was a categorical yes.[74]

In addition to the *nafaqa* system, the blind could receive assistance from certain charitable pious endowments. Many pious endowments provided food and shelter free of charge to the poor, and some reserved spots for blind people. For example, in sixteenth-century Cairo a large lodge run by a Sufi order housed 29 blind men out of 200, and in sixteenth- and seventeenth-century Damascus there was a *waqf* dedicated exclusively to assisting the blind (*waqf al-ʿumyān*).[75] In addition, the prestigious Al-Azhar University in Cairo has included for centuries a special college reserved exclusively for blind students.

73 Al-ʿImadi, [*Fatawa*], 1:200. Another instance of this consensus of opinion is found in Ibn ʿAbidin, *Radd al-Muhtar*, 6:579.

74 Al-Tarabulusi, *Fatawa*, 66a. See also Muhammad b. ʿUmar al-Hanuti (d. 1601/2), *Muntakhab al-Fatawa al-Hanuti* [*sic*], MS, Chester Beatty Library, Dublin, Ar. 5280, 84a–84b.

75 For Cairo, see ʿAbd al-Wahhab al-Shaʿrani (d. 1565), *Lata'if al-Minan wa-l-Akhlaq fi Wujub al-Tahadduth bi-Niʿmat Allah ʿala al-Itlaq* (Cairo: n.p., 1938), 2:158–160. For Damascus, see Ibn Ayyub, *al-Rawd al-ʿAtir*, 19; and El-Zawahreh, *Religious Endowments*, 258.

Some blind people, especially those without a familial support system, resorted to begging. Beggars could be resourceful and persistent, if not downright aggressive. Blind beggars, as well as beggars with other impairments, were notorious in many Arab cities, especially in the metropolis of Cairo.[76] Number 27 on a list compiled by the Turkish traveler Mustafa 'Ali (d. 1600) that one might well dub "Fifty Things I Hate About Cairo" reads as follows:

> The importunity of their beggars, the multitude of blind mendicants, the abundance of pestering paupers, lacking a member or paralyzed, one-handed or one-footed, showing various diseases, or – far may it be from those who listen [to me], far may it be from sight! – with grave wounds or sores, reaches such proportions that for one *manqur* they invoke God's name a thousand times and swear oaths. As soon as you put your hand in [your] pocket, you are surrounded by them, they all flock to you.[77]

Blind mendicants continued to importune the inhabitants of Cairo for centuries to come. In the late eighteenth century, on the eve of Napoleon's occupation of Egypt, the members of Al-Azhar's college for the blind were known for their collective and well-organized militancy. The Egyptian historian 'Abd al-Rahman al-Jabarti reports that the head of the college was an astute and powerful leader who would send "the gangs of the blind" to the tax-farmers who refused to pay him. These gangs were apparently so combative that debtors were intimidated into paying. The blind men of Al-Azhar "sustained themselves [...] on what they collected from begging on their nightly rounds and during the day in the markets and alleys, and from their singing eulogies and tales and reciting the Koran in houses and on the stone benches in the streets and so forth." Some of them, al-Jabarti concludes with a hint of bitterness, were actually quite wealthy.[78] This trope of

76 Between the thirteenth and the fifteenth centuries several Mamluk rulers endeavored to expel from Cairo some groups of disabled beggars, especially lepers and cripples (but not blind beggars). For reasons that remain unclear, a village in which their needs would be taken care of had been reserved for them in large oasis of Fayyum. All attempts were unsuccessful and short-lived. See Adam Sabra, *Poverty and Charity in Medieval Islam: Mamluk Egypt, 1250–1517* (Cambridge: Cambridge University Press, 2000), 60–61. I have not encountered evidence of similar efforts in the Ottoman period.

77 Tietze, *Mustafa 'Ali's Description of Cairo of 1599*, 49. A *manqur* was a coin of small value.

78 In Jane Hathaway, ed., *Al-Jabarti's History of Egypt* (Princeton, NJ: Markus Wiener Publishers, 2009), 196.

the greedy blind beggar, a familiar one in other cultural contexts as well, probably had its roots in sighted people's anxieties about the blind feigning their impairment and earning money without having to work.[79]

Evidently, begging paid off. This was due in large part to the charitable imperative that was an essential component of Islamic piety. At least in theory, for it is difficult to ascertain how often this rule was enforced, Muslims who could afford to do so were required to pay a yearly tax called *zakat* to be given as charity to the needy. The tax, which corresponds to 2.5 percent of a person's assets, is one of the "Five Pillars of Islam." In addition to this tax, Muslims were also strongly encouraged to give voluntary alms (*ṣadaqa*) to the poor.[80]

The story that follows illustrates the importance of public displays of benevolence and just how deeply ingrained in society this charitable imperative was. Alexander Russell, the physician who served the English colony in Aleppo in the eighteenth century, reported the following incident, which he heard from the mufti of Aleppo. One Friday, en route to the communal prayer, the mufti noticed among a group of beggars an old blind man who appeared to be quite ill. In a burst of perhaps excessive generosity, the mufti offered the old man a silver coin – many times the sum usually donated to a single beggar. For the following two years the mufti felt a moral obligation to give the blind man the same sum weekly, although, as he confessed to Dr. Russell, he had never expected the old man to live so long, and had thus occasionally "been disposed to repent having exceeded the ordinary bounds of his charity." Eventually the mufti noticed that the old man was gone, his usual spot now occupied by a much younger and only partially blind beggar. The mufti gave the replacement a copper coin, which prompted aggressive demands for a silver coin from the beggar, who protested that he had "acquired the station from the blind Sheikh, who is no longer able to come there." Unable to persuade the obstinate beggar that his younger age and blindness in only one eye did not

79 For the trope of greediness in France during the Middle Ages see Wheatley, *Stumbling Blocks*, especially pp. 22 and 64.

80 For more on charity, see Michael Bonner, Mine Ener, and Amy Singer, eds., *Poverty and Charity in Middle Eastern Contexts* (Albany: State University of New York Press, 2003), and Sabra, *Poverty and* Charity. Medical benevolence is discussed in Shefer-Mossensohn, *Ottoman Medicine*, 101–144.

warrant the same level of charity bestowed upon the older and completely blind man, the mufti had to content himself with a tentative promise that the beggar would not in turn sell his spot to someone else when he was no longer able to occupy it.[81]

This story confirms the centrality of charity to the identity of Muslims, especially prosperous ones. This was doubly the case for someone like Russell's informant who, as a mufti, had a prominent and public position and an accompanying social identity, the maintenance of which required repeated enactment. Herein lies the explanation for the mufti's behavior when confronted with the younger man's demands for a silver coin. Despite his palpable longing to "repent" from surpassing "the ordinary bounds of charity," he eschewed a decisive break with the importunate beggar, for doing so could have borne negative consequences for his reputation. Showing parsimony in such a symbol-laden setting, outside a city mosque before the Friday prayer, might inspire gossip that would cast doubt upon the mufti's charitableness, and, by extension, his piety. The blind Sheikh and his more able-bodied successor clearly understood this well. The tale demonstrates that even the most vulnerable blind Muslims possessed a sophisticated appreciation of the cultural norms within which they were embedded and could exercise considerable ingenuity in attempting to negotiate their way through them. It also suggests that blindness and other physical impairments may have at times actually improved the chances of the desperately poor, giving them a comparative advantage over their nonimpaired counterparts.

Conclusion

To return to the question posed at the beginning of this chapter, why is it improbable that spectacles of humiliation like the one narrated by the anonymous bourgeois of Paris in 1425 were ever staged in the Ottoman Arab world? On one level, the high prevalence of blindness resulting from trachoma (which France was spared) meant that its sheer pervasiveness in the early modern Arab-Islamic world normalized this sensory impairment, at least to some degree, thus making it somewhat less likely to be stigmatized.

[81] Russell, *The Natural History of Aleppo*, 1:205–206.

More fundamentally, the influence on later generations of Muslims of the decidedly favorable representations of blind people recorded in the foundational text of Islam cannot be overstated. Just as the Qur'an did not associate physical deafness and muteness with impiety or moral corruption, it did not associate physical blindness (as opposed to metaphorical blindness) with ritual impurity or deficiency of spiritual awareness, nor did it suggest that visual impairment be understood as the product of sin or the result of divine punishment. Equally paramount was the Qur'an and Hadith's complimentary depiction of 'Abd Allah Ibn Umm Maktum, the blind companion of the Prophet and one of the first muezzins and prayer leaders of the nascent Muslim community.

These positive ideas about blindness were incorporated into Islamic law, which, both in theory and in practice, strove to include and make accommodations for blind people in most spheres of life, the notable exception being the presentation of testimony in court proceedings. In addition, the powerful religious and cultural imperative of charity conveyed through the channels of *nafaqa* (maintenance by the family), *waqf*s (pious endowments), *zakat* (almsgiving tax), and *ṣadaqa* (voluntary charity) provided a safety net for the visually impaired.

Another crucial reason the marginalization of blind men was restricted is that many were gainfully employed in a variety of occupations, including prestigious ones, and that the highly educated and accomplished blind scholar and teacher was not an uncommon figure in the Ottoman Arab world. These achievements were made possible by the important role of orality in the educational system and in the transmission of knowledge and by the centrality of verbal performance in the recitation of the Qur'an, which did not necessarily privilege the sighted.

All of the above is not to suggest that people with visual impairments faced no or few obstacles, for they most certainly did. Blindness was doubtlessly considered an undesirable condition by mainstream society and a source of distress for many who experienced it. And, as fatwas from the period reveal, at times the husbands, wives, and relatives of blind individuals attempted to exploit the impairment to their advantage. Furthermore, socioeconomic status was critical in determining the educational and employment opportunities available to blind people. But ultimately, even though poor blind beggars were often viewed with suspicion the "amusements" described by the anonymous bourgeois of Paris were not staged because they were predicated on a pervasive negative characterization of the blind as vile and dim-witted and on a process of "othering" that, for the reasons explained in this chapter, did not take root in Arab-Islamic world.

3

Impairments of the Mind

"Madness Takes Many Forms"

It would be futile to engage in retrospective diagnosis or attempt to gauge the prevalence of specific impairments of the mind in the early modern Arab-Islamic world (or anywhere else in premodern times, for that matter). We can safely assume that some age-related degenerative diseases of the brain, such as Alzheimer's disease, which today is the most common form of progressive dementia and affects mostly people over the age of sixty-five, must have been far less common at a time when the average life expectancy was probably only about forty years.[1]

We also know that in the Middle Ages and early modern times parts of Europe were swept by harrowing epidemics of ergotism, also known as "St. Anthony's Fire" or "Holy Fire," a disease that caused excruciatingly painful physical and mental symptoms: burning sensations and gangrene of the extremities, convulsions, epileptic-like fits, hallucinations, delirium, and stupor. These epidemics, which some historians have suggested may have played a role in mass-scale accusations of witchcraft in Europe, were virtually unknown in Greater Syria and Egypt. The reason is simple: ergot poisoning is caused by eating cereals, predominantly rye, that have been contaminated by the ergot fungus, and rye was not commonly consumed in the Arab Middle East,

[1] Shefer-Mossensohn, *Ottoman Medicine*, 11.

where the climatic conditions were unsuitable for the growth of the fungus.

It is plausible that other types of mental disturbances were more common than they are today. For instance, dementia is one of the main symptoms of pellagra, which is caused by a chronic deficiency of niacin (vitamin B3). Pellagra was endemic among poor peasants living in the Egyptian Nile Delta, whose diets relied predominantly on corn, the only grain that lacks digestible niacin. Parts of Greater Syria, on the other hand, suffered from severe iodine deficiency, a common cause of problems with brain development and cognitive impairments.

But above all, there would have been scores of mental disturbances that resulted from accidents, that were congenital, or hereditary – the latter in no small part due to the high prevalence of consanguineous marriages in the region.

Early modern Arabs divided impairments of the mind into four broad groups: idiocy, melancholia, madness, and holy folly. The least fluid and ambiguous category was idiocy (*'atāha*), also known in English in the past as "mental retardation" or "imbecility" and today commonly referred to as cognitive or intellectual disability. Idiocy was viewed as a permanent state and it was almost always congenital or acquired in early childhood. It will not be explored in this chapter because, although it is discussed in Islamic law along with other impairments of the mind, it only rarely appears in sources other than legal writings and thus I was able to find only a handful of portraits of people labeled "idiots." The scarcity of "idiots" may be due precisely to the fact that idiocy was perceived as less complicated or less worthy of investigation: it was believed to be incurable, unlike melancholia and madness, and it lacked a connection with the divine, unlike holy folly.

Melancholia, madness, and holy folly, on which this chapter will focus, were more complex and protean categories. The amount of information on these conditions contained in Ottoman-era sources is enormous. Wading through this embarrassment of riches, one is compelled to acknowledge the truth of the age-old Arabic aphorism "*al-junūn funūn*," "madness takes many forms." Attempting to uncover how the people of Ottoman Syria and Egypt understood impairments of the mind is perhaps even more daunting than trying to map how deafness and blindness were conceptualized and experienced.

Impairments of the mind were everywhere, and it is difficult to overestimate their importance when "every important aspect of a Muslim's life was dependent on his being sane."[2]

Black Bile and the Porous Boundaries between Melancholia and Madness

On a Saturday in late May of 1591, the Damascene judge and historian Ibn Ayyub al-Ansari paid a visit to his friend Muhammad al-Salihi, who had recently experienced an unspecified tribulation so vexing that he had fallen ill and shut himself away in his home. When Ibn Ayyub entered Muhammad's room he found his friend sprawled on his bed as if he had been flung there. The ensuing conversation began as follows:

- Ibn Ayyub: "How is your temperament (*mizāj*)?"
- Muhammad: "Malignant (*khabīth*)."
- Ibn Ayyub: "How is your constitution (*ṭabī'a*)?"
- Muhammad: "Polluted (*najisa*)."
- Ibn Ayyub: "What are you eating?"
- Muhammad: "Dog meat."

After this exchange, Ibn Ayyub recited the Fatiha, the opening Sura of the Qur'an, then rose and departed. He later recorded his assessment of his friend's state: "I suppose that black bile (*al-sawdā'*) had overcome him, resulting in dryness and burning in his head. [. . .] There is no power and no strength save in God."[3] This Muhammad al-Salihi served as a notarial witness in one of the courts of Damascus and had acquired some acclaim, as well as numerous enemies, as the irreverent author of satirical poetry.[4] His droll repartee with Ibn Ayyub must thus be understood, at least to some degree, as a manifestation of poetic wit. The telegraphic economy of words with which Muhammad

[2] Dols, *Majnūn*, 4.

[3] Sharaf al-Din Musa Ibn Ayyub al-Ansari (d. ca. 1592), *Nuzhat al-Khatir wa-Bahjat al-Nazir*, ed. 'Adnan Muhammad Ibrahim (Damascus: Manshurat Wizarat al-Thaqafa, 1991), 2:198–199.

[4] The biography of Muhammad al-Salihi al-Hilali (d. 1596) is recorded in al-Ghazzi, *Lutf al-Samar*, 1:131–133; al-Burini, *Tarajim al-A'yan*, 2:82–83; and al-Muhibbi, *Khulasat al-Athar*, 4:34–36.

responded to his friend's solicitous inquiries seems to have come close to exasperating his visitor. These terse replies were obviously intended to convey, through metaphor and allusion, the abjectness of the poet's state, for "polluted" refers to ritual impurity, and dog meat was considered filthy and disgusting. Most importantly, this exchange encapsulates how many early modern people understood the inner workings of their bodies and conceptualized physical and mental illnesses.

As previously mentioned, according to humoral theory a person's temperament (*mizāj*) at any given time was dictated by the relative proportions of the four humors (blood, yellow bile, black bile, and phlegm) present in his or her body. Each of the four humors was associated with one of the four basic elements (air, fire, earth, water), with qualities (dry, wet, hot, cold), with the seasons, with different organs of the body, and with temperaments.

	Blood	**Yellow Bile**	**Black Bile**	**Phlegm**
Temperament	Sanguine	Choleric	Melancholic	Phlegmatic
Element	Air	Fire	Earth	Water
Season	Spring	Summer	Fall	Winter
Qualities	Wet and Hot	Dry and Hot	Dry and Cold	Wet and Cold
Organ	Heart	Liver	Spleen	Brain

Viewed within this framework, Ibn Ayyub's initial query about the temperament of his ailing friend signified more than a generic interest in his condition; it also made perfect sense from a diagnostic point of view. Ibn Ayyub's postvisit comment that melancholia had overpowered Muhammad and that his head was affected by "dryness and burning" reflected, albeit in a distilled and simplified form, medical theory of the time.

Exactly what was this black bile of which Ibn Ayyub spoke? Physicians devoted lengthy sections of their medical compendia to the "ailments of the head." Under this heading they catalogued many distinct forms of headache and a wide range of conditions including memory loss, insomnia, lethargy, inflammation of the brain (*sirsām*), hemiplegia (total or partial paralysis of one side of the body), nightmares, epilepsy, apoplexy, vertigo, and melancholia (*mālīkhūliyā* or *mālankhūliyā*). On account of being subsumed under the category of "ailments of the head," all these illnesses were, quite literally, "mental illnesses."[5]

The last of these ailments, melancholia, is the one that is most frequently discussed. The term *mālīkhūliyā* was an Arabic adaptation from the Greek "melancholia," which means "black bile." al-Antaki (d. 1599) prefaced his discussion of melancholia by stating that the term described "many forms that differ a little according to the symptoms," but all forms of the illness were derived primarily from one source, "the corruption of the brain and the mind due to an excess of the two dry humors," that is, yellow and black bile. According to al-Antaki, paradigmatic cases of melancholia were caused by "the filling up of the entire body with bile," resulting in the symptoms of confused thought processes, irritability, distrust, suspicion, and delusions.[6] Ibn Sallum (d. 1670/1), too, attributed the disease to "an excess, putrefaction, or burning of the humors," and listed its symptoms as "sleeplessness, nightmares, a tendency to become agitated with little or no cause, anger, fear, love of solitude, withdrawal from human company, and distrust," among others.[7]

Physicians concurred that melancholia actually comprised a group of illnesses. One "melancholic illness" (*maraḍ sawdāwī*) was hypochondriacal melancholia. The term "hypochondriacal" refers to the upper abdomen. In this type of melancholia, the putrefaction of the humors occurred in the stomach, spleen, and liver and ascended from there to the head. The symptoms of hypochondriacal melancholia included the "love of solitude," taciturnity, a powerful sense of physical fragility (literally, "imagining that one is a glass that shatters"), "being certain of what is not there, such as that someone wants to kill [you]," delirium, fear, and "anger triggered by the most trifling thing." Other forms of melancholia were *māniyā*, also known as the "bestial disease" or the "bestial madness," because those afflicted behaved like dogs or other predatory animals, delirium (*hadhayān*),

[5] For full lists of "ailments of the head" (*amrāḍ al-ra's*), see al-Antaki, *al-Nuzha al-Mubhija*, 2:96–132; Ibn Sallum, *Ghayat al-Itqan*, 1b–42a. Al-Antaki and Ibn Sallum include slightly different illnesses. Unlike al-Antaki, for instance, Ibn Sallum did not include passionate love (*'ishq*) and lycanthropy (*quṭrub*) in his treatise.

[6] Al-Antaki, *al-Nuzha al-Mubhija*, 2:110. The same description also appears in al-Antaki, *Tadhkirat Uli al-Albab*, 3:165.

[7] Ibn Sallum, *Ghayat al-Itqan*, 28a.

and *junūn* (madness).[8] The last, *junūn*, was classified as one of the more violent forms of melancholia and is the most common Arabic word for madness to this day. Al-Antaki offers the following definition of *junūn*: it is "the departure or impairment of the mind," a state that "can be either continuous or intermittent."[9]

For the treatment of melancholia physicians prescribed bloodletting, purges, the strict adherence to a diet consisting of chicken, milk, eggs, lettuce, and a handful of medicinal potions and recipes. These remedies were thought to reduce the proportion of black bile in the patient's system. Listening to music could also provide therapeutic benefits. Indeed, music therapy enjoyed a long history in Arab-Islamic medicine, and in the Ottoman period musicians gave regular performances in the hospitals of Cairo and Damascus.[10]

Medical theories about mental disturbances may be summarized as follows. Physicians posited a close relationship between melancholia and *junūn* and made a quantitative rather than a qualitative distinction between the two conditions. The surplus or putrefaction of black bile in the body was thought to result in a melancholic (*sawdāwī*) temperament. Depending on the location of the black bile's accumulation, on the underlying cause of this accumulation or corruption, and on the bile's interaction with other humors, particularly yellow bile, the other "dry" humor, the melancholic's behavior could be sad, despondent, irascible, aggressive, violent, or delusional.

With this background in mind, let us return to the repartee between Ibn Ayyub and his ailing friend. The humoral framework for understanding melancholia clearly informed Ibn Ayyub's comments about Muhammad al-Salihi's mental state. Black bile had gotten the better of him and "dryness and burning resulted in his head," a reference to the dry nature of black bile and to the substance's corruption or putrefaction, its "burning." Ibn Ayyub's inquiry about his friend's diet was

8 Al-Antaki, *al-Nuzha al-Mubhija*, 2:111–112; Ibn Sallum, *Ghayat al-Itqan*, 29b, 31a–31b.

9 Al-Antaki, *Tadhkirat Uli al-Albab*, 2:84–85. Four lengthy chapters of Michael Dols's *Majnūn* explore the evolving understanding of the relationship between humoral imbalance and madness and the treatment of the latter from the Greeks to the medieval Islamic Middle East: "Galen and Mental Illness," "Galen into Arabic," "The Reformulation of Greek Medicine," and "Medical Madness," 17–103.

10 Al-Antaki, *al-Nuzha al-Mubhija*, 2:112–115; Ibn Sallum, *Ghayat al-Itqan*, 28a–29b. For music as medical therapy, see Shefer-Mossensohn, *Ottoman Medicine*, 69–77.

in full conformance with humoral theory, too. Certain foods, especially those believed to enable or increase the production of black bile (lentils, beef, and cabbage, for example) ought to be avoided by those experiencing melancholia.

There is no reason to assume that this exchange between two friends, neither of them a medical professional, was in any way extraordinary. To the contrary, physicians and nonphysicians alike shared the same basic conception of the body, its modes of functioning, and its relationship to the mind and to nature. To be sure, *mālīkhūliyā* was a technical term used primarily by physicians and thus is almost never found outside of medical texts. However, expressions like "black bile overpowered him," "black bile afflicted him," or "a melancholic condition happened to him" recur very frequently in narratives of mental disorders recorded in nonmedical sources. These sources include dream manuals, where nightmares were ascribed to the predominance of black bile in the body.[11]

On occasion, melancholia was attributed to an individual's character or constitution. An eighteenth-century mufti, for example, was described as someone who "shunned people and whose nature was not devoid of melancholia (*sawdā'*)."[12] Usually, however, it was conceived as a condition that *happened to* someone, as the outcome of a process, the effect of an underlying cause or series of causes.

There was general agreement among laypersons and physicians that black bile was the physiological agent that engendered melancholia, but there was little agreement on anything else. The etiologies of melancholia were many and diverse and the forms it took were multifarious. In addition to medicine, remedies included talismans, amulets, incantations, and the water of special springs like "the Spring of Life," in present-day south Lebanon.[13]

Certain manifestations of melancholia remind us of what is currently called "depression." The melancholic are described as recluses

[11] Al-Ghazzi, *Luft al-Samar*, 1:36, 275; al-Burini, *Tarajim al-A'yan*, 1:268; al-Muhibbi, *Khulasat al-Athar*, 4:367; al-Muradi, *Silk al-Durar*, 2:125, 3:38, 4:241; al-Nabulusi, *Ta'tir al-Anam*, 1:3.

[12] Al-Muradi, *Silk al-Durar*, 1:275.

[13] For talismans and amulets, see Dols, *Majnūn*, 235–236. The "Spring of Life," which was "effective against all sorts of illnesses affecting the brain, as well as melancholy, haemorrhoids, and white morphew," is described by Evliya Çelebi in *Evliya Tshelebi's Travels in Palestine (1648–1650)*, 13.

who insulate themselves from contact with friends, neighbors, and colleagues. Some displayed delusional tendencies. The talented Ahmad al-Kaywani (d. 1759), who was regarded by many of his peers as the foremost Syrian poet of his century, struggled for some time with a severe case of melancholia. He tended to shun the company of fellow humans, was reputed to imagine unspecified "strange things," and was known to "lament the age in which he lived" so frequently that a fellow poet once characterized the essence of al-Kaywani's opus as nothing more than moaning and whining.[14] Incidentally, al-Kaywani means "Saturnine." This is unlikely to be a coincidence, for the planet Saturn, regarded as the coldest and driest of planets, was linked to melancholy.[15]

Melancholia associated with sadness and despair was often attributed to environmental factors. Its onset was triggered by the loss of parents, spouses, or children, events that could understandably plunge an individual into deep sadness. Powerful emotions like grief, ardent love, fear, and anger, whatever their external cause, were understood as physiological phenomena, and therefore intense feelings and emotions, if ungoverned, were thought to have a direct and deleterious impact on humoral balance. For example, physicians described the physiological processes resulting from anger in the following terms: "Anger is the boiling of the blood in the heart and the resulting transfer of natural heat throughout the rest of the body." Its harmful effects, physicians believed, were manifested by "redness in the eyes and a flushed face."[16]

Not all melancholic individuals presented symptoms that are easily relatable to our contemporary notions of depression. Mustafa al-Muhibbi (d. 1651) began to experience severe confusion in the middle years of his life, a condition that was ascribed to the prevalence of burned black bile in his system. Because of this confusion he started

[14] Al-Kaywani's biography as well as several excerpts from his poetic oeuvre appear in al-Muradi, *Silk al-Durar*, 4:113–124. 'Umar Musa Basha dedicates a lengthy section of his history of Arabic literature in the Ottoman period to al-Kaywani's work. See *Tarikh al-Adab al-'Arabi: Al-'Asr al-'Uthmani* (Beirut: Dar al-Fikr al-Mu'asir; Damascus: Dar al-Fikr, 1999), 506–565. For another case of delusion and paranoia, see the biography of the great-uncle of the biographer and mufti Muhammad Khalil al-Muradi in al-Muradi, *Silk al-Durar*, 2:122–126.

[15] The connection between astrology and physical and mental impairments in the Arab-Islamic world remains to be explored.

[16] Cited in Dols, *Majnūn*, 64. The physician in question is 'Ali al-Majusi (d. 994).

to act strangely. He shouted disparaging comments about prominent scholars of his time from the minaret of a mosque, and would write confused or irrelevant notes (at least in the estimation of his peers) in the margins of his books and even of those he had borrowed from colleagues – evidently a breach of etiquette. He spent the last years of his life secluded in his home.[17]

Shihab al-Din al-Ghazzi was a much older half-brother of the famed biographer Najm al-Din al-Ghazzi. Like many male members of the Ghazzi family in Damascus, Shihab al-Din distinguished himself in several fields of scholarly and spiritual endeavor. He served as mufti, teacher, and imam, and later became a prominent Sufi. Shihab al-Din was also noted for the extraordinarily close and affectionate relationship he enjoyed with his father, the Shafiʿi mufti Badr al-Din al-Ghazzi, a bond that is described in touching detail. Shihab al-Din was very sad for much of his life and suffered from frequent headaches of such intensity that he was said to moan aloud in pain. He also lost his appetite for about two years and became emaciated as a result. He could find no respite from *waswasa* (a type of severe melancholy resulting in anxiety and delusions), was always on his guard, and developed an obsession with ritual purity. So severe and distinctive were these symptoms that Shihab al-Din's own brother described them in terms that might meet today's criteria for a diagnosis of paranoia and obsessive-compulsive disorder. Ironically, Shihab al-Din died in the pursuit of his obsession with cleanliness, following his daily habit of performing his preprayer ablutions at a public bath. He collapsed and expired there at dawn in early March of 1576.[18]

Many other stories confirm that the conceptual boundaries between madness and less aggressive forms of melancholia were elastic and permeable. Sometimes, indeed, the two terms were used interchangeably. Abu al-Tayyib al-Ghazzi, yet another brother of Najm al-Din al-Ghazzi, is a good example. A close friend of Abu al-Tayyib produced a sympathetic account of his melancholic ailment and its terrifying consequences. In 1606 or 1607, Abu al-Tayyib divorced his wife. Soon thereafter he began to mutter dark threats against his young

[17] Al-Muhibbi, *Khulasat al-Athar*, 4:367.

[18] Biographical information on Shihab al-Din al-Ghazzi is drawn from Ibn Ayyub, *al-Rawd al-ʿAtir*, 13–18, and from al-Ghazzi, *al-Kawakib al-Saʾira*, 3:100–109.

son, saying that the boy must be killed before achieving maturity or else something terrible or shameful would happen to him. The young boy had to be shielded from Abu al-Tayyib, who was by now considered sufficiently dangerous to warrant confinement in his father's home. Despite his lamentable state, Abu al-Tayyib continued to compose excellent poetry, for which he gained considerable acclaim, and to pursue other intellectual interests until his death. One biographer did not hesitate to classify Abu al-Tayyib's ailment as sheer insanity (*junūn*), but others preferred to use the more understated phrase "overcome by humors."[19]

The ambiguous boundaries between melancholia and madness are perhaps best exemplified by the life of Ahmad Ibn al-Minqar (d. 1623). Born in Damascus into a prominent family, he showed extraordinary promise early in life and wrote a treatise on the use of metaphors in literature before the age of twenty. He went on to become a well-respected teacher and poet. In his late thirties Ibn al-Minqar traveled to Istanbul, where he joined the entourage of the Grand Mufti. In the midst of this success, "black bile overpowered him" and his mind became "muddled and defective." The ultimate cause of Ibn al-Minqar's misfortune was unclear, but it might have been connected to the fact that while in Istanbul he began to consume unspecified drugs, a practice that was said to cause the onset of melancholia. His speech became confused and his actions so erratic that he was taken to a local hospital, where his condition resisted treatment. Ibn al-Minqar's mental state deteriorated so severely that some of his Damascene acquaintances were required to accompany him back to his hometown, a journey that the scholar made in chains. One biographer, who was on friendly terms with the disturbed man, reports being deeply distressed by the sight of his arrival in Damascus. Ibn al-Minqar's family subsequently tried every known treatment for his condition, but all to no avail. His madness (*junūn*) was intermittent, the severity of his condition tending to increase or decrease depending on the season of the year. This last observation, the apparent ebb and flow of symptoms according to seasonal rhythms, is again in conformity with humoral theory, which associated each of the four humors with a particular season. When his symptoms were in

[19] Al-Burini, *Tarajim al-A'yan*, 1:268–269; al-Muhibbi, *Khulasat al-Athar*, 1:135.

remission he behaved relatively normally for brief periods of time, but during more prolonged interactions the confusion (*ikhtilāl*) of his thoughts, speech, and behavior would become obvious to all. Because of the intractability of his condition, Ibn al-Minqar was almost continuously confined to his home for about thirty years. Under the supervision and care of a custodian hired by his family, he remained there until his death, enjoying only brief and occasional outings.[20]

Home confinement was the standard practice for those who suffered from severe and persistent forms of "mental disturbance" (*ikhtilāl al-'aql*) and whose families owned sufficient means to ensure that they did not harm themselves or others.[21] If there was no family safety net or if the family was unwilling or unable to care for a mad person, after a judicial order was issued he or she would be taken forcibly to a hospital for treatment and would have to remain there until cured, at the expense of the state. Hospitals were open to anyone regardless of gender, class, and religion, but in practice they were a last resort that served the needs of the poor, travelers, and the violently mad.

In the Ottoman Arab world there was no "Great Confinement" like the one described (perhaps simplistically) by Michel Foucault for seventeenth- and eighteenth-century France. Foucault argued that this "Great Confinement," which took place in the context of the Enlightenment and of political absolutism, resulted in the systematic institutionalization of members of society viewed as nonproductive or disruptive. These undesirables included not only the insane but also the poor, vagrants, heretics, single mothers, and failed suicides, among others.[22] At no point during the early modern period was institutionalization widespread in Ottoman Syria or Egypt. Evliya Çelebi recounts that when he visited Cairo in the seventeenth century there were only four hospitals in the city, the largest of which, the Mansuri Hospital, catered

[20] Al-Burini, *Tarajim al-A'yan*, 1:163–177; al-Ghazzi, *Lutf al-Samar*, 1:274–275; al-Muhibbi, *Khulasat al-Athar*, 1:296–297.

[21] Another example is Muhammad al-Qudsi (d. 1599/1600). When he ceased to recognize friends he saw in the street and began to exhibit other odd and inexplicable tendencies, his behavior was attributed to the onset of *ikhtilāl*. As a result, his son confined him to their home. Al-Burini, *Tarajim al-A'yan*, 3:158b.

[22] Michel Foucault, *History of Madness*, ed. Jean Khalfa and translated from French by Jonathan Murphy and Jean Khalfa (London; New York: Routledge, 2006).

to a combined total of 306 ill and insane patients.[23] Even if we accept his figure at face value (Çelebi is notorious for his exaggerations), it would still be a negligible number for a metropolis like Cairo. Furthermore, hospitals were usually centrally located, underscoring that their goal was not to hide or insulate inmates from the rest of the population.

The hospitals of Cairo, Damascus, and Aleppo were impressive buildings with pools, fountains, and gardens, where music therapy was sometimes employed, but the insane wards of Ottoman-era hospitals must not be romanticized. Hospitals were viewed as fearful places: if they appeared in a dream, they indicated death and hell, "because the hospital is the place of chains and shackles."[24] Çelebi reports that in Cairo men and women deemed dangerous were "bound like lions with chains around their necks," and doctors had "fearless servants with the temperaments of executioners," who beat the insane people with cherry switches until they "came to their senses."[25]

Spirit Possession

Melancholia, in its various forms that included withdrawn moroseness, mental confusion, and violence, was caused by a change for the worse in the temperament of an individual. This change often involved a surplus or putrefaction of black bile. A number of agents were posited as the cause of this underlying physiological change: emotions, especially grief and love, certain foods, the seasons, a natural predisposition, the consumption of drugs, and occasionally sorcery. In one instance, a man's reason was "snatched away" by an unspecified substance that his foes intentionally put in his food.[26]

The jinn, however, were only rarely invoked explicitly as the agents of insanity. Given that the word *majnūn* literally means "possessed

[23] Gary Leiser and Michael Dols, "Evliyā Chelebi's Description of Medicine in Seventeenth-Century Egypt. Part II: Text," *Sudhoffs Archiv* 72, no. 1 (1988): 53. For a fuller discussion of hospitals in the Ottoman period, see Shefer-Mossensohn, *Ottoman Medicine*, especially 110–144.

[24] Al-Nabulusi, *Ta'tir al-Anam*, 2:233.

[25] Leiser and Dols, "Evliyā Chelebi," 53–54.

[26] Abu al-Wafa' al-Halabi al-'Urdi (d. 1660), *Ma'adin al-Dhahab fi al-A'yan al-Musharrafa bi-him Halab*, ed. Muhammad Altunji (Damascus: Dar al-Mallah, 1987), 71.

by a jinni," this absence is perplexing. The jinn, also translated into English as "spirits," "genies," or "demons," are believed to be supernatural and usually invisible beings cohabiting the earth with humans, in whose lives they can intervene to beneficent or malevolent effect. Their existence is attested to in the Qur'an and they are a topic of discussion in Islamic law, particularly in reference to their marriages to humans, which are precisely regulated. But their omission from discussions of the etiology of madness raises the question of what evidence there is that the people of Ottoman Egypt and Syria actually believed in the power of these supernatural beings.

The jinn do not appear frequently in early modern Arabic writings. Ibn Tulun (d. 1546) reports two incidents involving the supernatural creatures. In the first, a jinni warned a woman in her sleep that "those who don't color with henna will suffer direly." The news quickly spread through Damascus, causing a dramatic increase in henna purchases and resulting in a shortage of the dye. The second incident, mentioned in Chapter 1, was that of the unfortunate man struck deaf by the blow of a jinni while answering a call of nature.[27]

Whatever their ostensible intentions, the jinn were reputed to be unpredictable and vindictive and consequently provoked fear. For example, the bedtime of judge Muhammad al-Sururi (d. 1678) was often disturbed by jinn, who used these nocturnal visits to "acquire knowledge from him." Although they never harmed him, these visits would fill al-Sururi with considerable anxiety.[28]

An unusually extensive account of a jinni appears in the biography of Khalil al-Sarmini, a baker's assistant in late sixteenth-century Aleppo. Al-Sarmini found himself at the center of an extraordinary story. During one of his many solitary nights spent kneading loaves of bread for baking, a group of jinn made their presence known. Among them was "a beautiful virgin" named Khatun, who liked him and wanted to marry him. The dough-kneader accepted the offer and the couple had a long and tumultuous marriage that became the talk of the neighborhood (she was reputed to be violently jealous of the pretty boys so admired by her husband). The couple was also reported to have had children, who, just like their mother, could be heard but

[27] In Ibn Tulun, *Mufakahat al-Khillan*, 1:105 and 1:259, respectively.

[28] Al-Muhibbi, *Khulasat al-Athar*, 3:414–415.

were invisible to humans. Al-Sarmini's biographer writes that he was a child when news of this very unusual marriage swept his neighborhood. He recalls seeing al-Sarmini at the mosque but admits that his fear of the jinn caused him to dread speaking with the man. The biographer's father, who was the chief Shafi'i mufti of Aleppo, did not share his son's terror of al-Sarmini and his unorthodox family. Perhaps in an attempt to get to the bottom of the matter, he invited the baker's assistant to his home for dinner and sought an interview with his invisible wife. The baker's assistant complied with the request, and the mufti satisfied himself as to the veracity of the story and the existence of the jinni wife.[29]

That the jinn were generally a source of apprehension, if not outright fear, is corroborated by dream manuals, which invariably associate their appearance in dreams with cunning, craftiness, magic, and danger.[30] Surprisingly, however, medical and biographical texts do not number them among the many possible causes of madness. The failure to posit a causal relationship between the two cannot be the result of individual or collective oversight. Da'ud al-Antaki, the luminary of Arab-Ottoman medical science, acknowledged the existence of these invisible and potentially dangerous beings, described their role in the etiology of other ailments, and devoted a lengthy section of his medical treatise to enumerating an array of spells, incantations, and talismans thought to offer protection against the jinn, the evil eye, and sorcery.[31] But neither he nor his contemporaries discussed them in the context of melancholia or madness. How can this apparent discrepancy be accounted for? Perhaps, as Michael Dols has suggested for the classical and medieval periods, the jinn "were usually spoken of by allusion and antiphrasis in order to avoid their powerful influence."[32] This may well have been the case in the Ottoman period as well, for many people undoubtedly believed in the existence of these potentially destructive creatures. That said, the available evidence supports only one conclusion: the early modern Arab-Islamic world was a cultural sphere in which supernatural forces were afforded only one of many

[29] This story is reported by al-'Urdi in *Ma'adin al-Dhahab*, 344–345.

[30] See al-Nabulusi, *Ta'tir al-Anam*, 1:105–106.

[31] Al-Antaki, *Tadhkirat Uli al-Albab*, 3:147–165.

[32] Dols, *Majnūn*, 214.

roles in the etiology of madness. Other explanatory models rooted in physiology and humoral theory had pride of place, at least in the literate world whose documentary traces have survived to us.

* * *

In sum, melancholia encompassed a broad spectrum of diverse symptoms, from profound sadness and a desire for seclusion to delusion, paranoia, and many behaviors deemed erratic, irrational, or dangerous. Its more aggressive or violent forms were called *junūn*. It was often challenging for laypersons and professionals alike to define, classify, and delineate the boundaries between different ailments of the head, but they were usually understood in physiological terms and were not deemed fundamentally different from ailments of the body stemming from natural and explicable processes. People were reluctant to ascribe mental disturbances to demonic possession or other supernatural forces, and even more rarely to God's punishment for the sins of the afflicted.[33] Quite the opposite: the most common form of madness in the Arab-Islamic world, holy folly, was actually understood as a sign of God's *favor*, for which cures were neither possible nor desired.

Holy Folly: A Different Kind of Madness

Evliya Çelebi, the celebrated traveler who spent about forty years of his life touring the Ottoman Empire and lands beyond, recorded his experiences in his voluminous *Book of Travels* (*Seyahatname*). In 1648 Evliya journeyed from Istanbul to Damascus with Murtada Pasha, the newly appointed Ottoman governor of the city. One day, during his yearlong residence in Damascus, Evliya was invited by ten Turkish officers to join them for a night of revelry in a house of ill repute. After hesitating, he (ostensibly reluctantly) accepted the invitation and joined the group on their nocturnal outing. En route to the establishment, Evliya suddenly came face to face with a local *majdhūb* (holy fool) by the name of Shaykh Bakkar. His description of the encounter is worth quoting at length:

> As the eleven of us fine gentlemen were sauntering through the crowd in the Sinaniye Market, we saw coming toward us the above-mentioned Shaikh

[33] This observation was also made by Michael Dols, *Majnūn*, 215.

Bekkar, stark naked, with his hands on his shoulders and his genitals swaying to and fro. Suddenly he came up to me and slapped my face with the slap of a holy idiot. I reeled, and blood gushed from my nose onto my green cloak. When I looked up, I saw that all my companions had fled. Shaikh Bekkar seized me by the hand and began to parade me around the marketplace as though he were a slave broker, shouting, "One sinner for sale! One new one! For sale! One new one!" He was actually auctioning me off! I was so humiliated that my entire body broke into goose-bumps. The crowd looked on amazed, and some boys even asked the shaikh how much he was selling me for. I was bathed in sweat from the shame of it.

He kept parading me about until we came to the convent of his holiness Shaikh Arslan. There he allowed me to renew my ablutions, after which we entered the shrine and he said: "Recite the Surah of the Emissaries." I recited it in a loud voice. "Recite the Surah of the Soul-Snatchers." I recited that as well. When I finished, he opened his hands and babbled a kind of nonsensical prayer. "This prayer nice, nice," he said, addressing me. He finished with: "Say, I ask forgiveness of God," and I replied: "Turn to God in repentance." Then he stood up and kissed my forehead, saying, "This tomb my master, you my child." Seizing me by the hand once again, he ushered me out of the convent. As we stepped outside, he kissed my shoes and placed them in front of me. I kissed his hand and put on my shoes, thinking, "What can I do? He is naked and crazy, and he has me in his clutches."[34]

The Shaykh then proceeded to drag Evliya through the marketplace once more, shouting again that the rascal was for sale, until they reached the governor's Palace. Once there, the Shaykh handed him to the Pasha. After kissing Evliya's forehead and proclaiming him "his spiritual son," Shaykh Bakkar left. Shortly afterwards, while Evliya was recounting his adventure with the Shaykh to the Pasha, their conversation was interrupted by the announcement that the ten gallivanting officers had been killed while resisting arrest at the brothel. The moral of the story, Evliya concluded, was:

I was on my way with these men to that den of vice. But because I am one of those who bear God's holy word, having memorized the Koran, his holiness Shaikh Bekkar received a divine inspiration and seized me from the midst of those doomed men. He paraded me round about, crying, "One sinner!" and

34 The translation of this passage is taken from Robert Dankoff, *An Ottoman Mentality: The World of Evliya Çelebi* (Leiden; Boston: Brill, 2004), 128–129.

so rescued me – may his secret be sanctified. Owing to that saintly man I was saved from that abyss.[35]

Even allowing for Evliya's characteristic dramatic embellishments, his description tallies with other available information about Shaykh Bakkar. We know that a popular holy fool by that name did in fact live in Damascus throughout the period of Evliya's residence in the city and died in 1665 or 1666. He was nicknamed "the naked one," he was known to have performed numerous miracles, and there was general agreement on his saintliness.[36] The sketch of Shaykh Bakkar penned by Evliya Çelebi nicely captures the gist of the *majdhūb* (pl. *majādhīb*): mad and saintly.

Holy fools were ubiquitous in Ottoman Syria and Egypt. Early modern Arabic writings teem with them, suggesting that the visual landscape of Arab towns and cities must have been dramatically different from today. Holy fools also left a vivid impression on Turkish and European travelers well into the nineteenth century. There was "an almost incalculable number" of them in Cairo, reported the Italian physician-botanist Prosper Alpin, who lived in Egypt from 1581 to 1584. "There are thousands of *majādhīb*," echoed Evliya Çelebi during his stay in Cairo in the seventeenth century. This assessment was shared by the Scottish physician Alexander Russell, who lived in Aleppo in the eighteenth century, and by a Spanish traveler (and possibly spy) in Damascus at the turn of the nineteenth century. The Spaniard wrote, with a tinge of contempt, that he observed "an infinity of individuals who call themselves saints, and appear as madmen or idiots, with a view to attract the veneration of the public."[37]

35 Dankoff, *An Ottoman Mentality*, 130. Some of Evliya Çelebi's descriptions are distinctly hyperbolic, a rhetorical device commonly used in travelogues and other genres of "entertainment" literature. See chapters 5 (Raconteur) and 6 (Reporter and Entertainer) in Dankoff's *An Ottoman Mentality*.

36 Al-Muhibbi, *Khulasat al-Athar*, 1:454–455.

37 Prosper Alpin (d. 1617), *La Médecine des Egyptiens par Prosper Alpin, 1581–1584*, edited and translated from Latin by R. de Fenoyl (Cairo: Institut Français d'Archéologie Orientale, 1980), 1:103; Evliya Çelebi, *al-Rihla ila Misr wa-l-Sudan wa-l-Habasha*, edited and translated from Turkish into Arabic by Muhammad Harb (Madinat Nasr, Cairo: Dar al-Afaq al-'Arabiyya, 2006), 2:83–84; Russell, *The Natural History of Aleppo*, 1:211; Ali Bey, *Travels of Ali Bey in Morocco, Tripoli, Cyprus, Egypt, Arabia, Syria, and Turkey. Between the Years 1803 and 1807. Written by*

Portraits of Holy Fools

Majdhūb, a word that has been variously rendered into English as "holy fool," "mystic," "ecstatic," "enraptured," "possessed," "crazed," or even, incorrectly, "Sufi," is a passive participle from the Arabic root j-dh-b, which means "to pull, to attract." It was implied and understood by all that the attraction was to God. Thus, a majdhub was someone who possessed "divine attraction," a state that could be acquired only through God's agency. This link with the divine was often reinforced with the inclusion of other religious terminology in descriptions of holy fools, such as *walī* (a person close to God, a saint), *kashf* or *mukāshafa* (illumination, unveiling) to signify the mystical unveiling of hidden divine truths, *baraka* (blessing), and *karāmāt* (miracles).

No explanations were proffered, nor were they expected, for God's choice to attract particular individuals and not others. Only rarely is the reader afforded a glimpse of the actual process of becoming attracted. Divine attraction could descend on anyone at any time and at any place and for no apparent reason. It "appeared," "seized," "happened to," or "overpowered" a person (the same verbs used, perhaps not coincidentally, to describe the onset of melancholia), usually an adult but occasionally also a child.[38] Some people experienced divine attraction only sporadically, but for most it was a constant, life-long state.

Many holy fools were recognizable by their physical appearance and by their clothing. Some shaved their heads, their beards, and at times even their eyebrows.[39] Since the beard was the sign of manhood *par excellence* in Ottoman-Arab society, habitually shaving one's facial hair created a dramatic physical marker that visually placed a man

Himself, and Illustrated with Maps and Numerous Plates. Second American from the First London Edition. (Philadelphia: Printed for John Conrad, at the Shakespeare Buildings, 1816), 2:231.

38 For examples of children who became holy fools as children, see Barakat Ibn al-Himsi (d. 1537/8) in al-Ghazzi, *al-Kawakib al-Sa'ira*, 2:127, and the young holy fool who accompanied 'Abd al-Ghani al-Nabulusi on his journey to Palestine in *al-Hadra al-Unsiyya fi al-Rihla al-Qudsiyya*, ed. Akram Hasan al-'Ulabi (Beirut: al-Masadir, 1990), 67.

39 Al-Ghazzi, *al-Kawakib al-Sa'ira*, 1:212, 213; 2:72; al-Ghazzi, *Lutf al-Samar*, 1:259, 306, 329; al-'Urdi, *Ma'adin al-Dhahab*, 46, 66, 180, 191, 192; al-Muradi, *Silk al-Durar*, 2:208; al-Muhibbi, *Khulasat al-Athar*, 4:511; 'Abd al-Rahman al-Jabarti,

outside the confines of adult manliness, thus constituting a symbolic and voluntary emasculation. On occasion, emasculation transcended the symbolic: some enraptured men actually castrated themselves.[40]

The refusal to wear a head covering, which was the standard and expected custom for men and women alike, and wandering the street naked or wearing only a loincloth were two other distinctive practices adopted by many, although by no means all, male holy fools. Some walked barefoot year round, indifferent to extremes of temperature.[41] Those who eschewed nudity sometimes wore filthy clothes. Sartorial eccentricity occasionally extended into the realm of "accessorizing," as in the case of a seventeenth-century majdhub who resided near Nablus and never appeared in public without a sword and a large firearm, or of ʿAli al-Bakri (d. 1792/3), who often walked the streets of Cairo naked but carrying a long club in his hand.[42] One famous Damascene holy fool in the sixteenth century was a corpulent man with a long beard and white hair. Far from wandering the streets naked, an instance of his singularity was that he draped himself in multiple layers of fur.[43]

Unconventional grooming and attire, however, should not be overstated as a distinguishing feature of holy fools. In several cases, authors, usually attentive to any remarkable aspect of the subjects about whom they write, make no mention at all of their appearances. Their silence may reasonably be taken as an indication that there was simply little remarkable about it. Looks, therefore, could not have been the primary marker of holy folly. Ultimately, what made a majdhub a majdhub was his actions.

A man, and more rarely a woman, and even more rarely a child, became a holy fool by publicly enacting the role with sufficient

ʿAja'ib al-Athar fi al-Tarajim wa-l-Akhbar, ed. Hasan Muhammad Jawhar, ʿUmar al-Dasuqi, and al-Sayyid Ibrahim Salim (Cairo: Lajnat al-Bayan al-ʿArabi, 1959–1967), 3:322.

40 Al-Shaʿrani, *al-Tabaqat al-Kubra*, 2:252. Another example of self-castration is reported in al-ʿUrdi, *Maʿadin al-Dhahab*, 259.

41 See, for instance, al-Shaʿrani, *al-Tabaqat al-Kubra*, 2:253, 254; Ibn al-Hanbali, *Durr al-Habab*, 2:498; al-Ghazzi, *al-Kawakib al-Sa'ira*, 1:78, 2:72; al-Ghazzi, *Lutf al-Samar*, 1:259, 1:306, 2:206; al-Nabulusi, *al-Hadra al-Unsiyya*, 327; al-Jabarti, *ʿAja'ib al-Athar*, 3:322; Çelebi, *al-Rihla*, 2:83–84.

42 Al-Nabulusi, *al-Hadra al-Unsiyya*, 82; al-Jabarti, *ʿAja'ib al-Athar*, 4:237.

43 Al-Burini, *Tarajim al-Aʿyan*, 3:163a.

authenticity and consistency to generate a critical mass of belief in his or her genuine saintliness. A large yet indeterminate number of people had to be convinced that a given individual's dress, speech, demeanor, and acts were congruent with their template of a "real" majdhub, that is, that he or she had demonstrated special abilities that could derive only from a profound and intimate connection with God.

An important criterion adopted to determine the genuineness of a holy fool was his ability to perform *karāmāt* (miracles, acts of divine grace) or *khawāriq* (wondrous deeds), often interchangeable terms that encompassed a wide range of feats. Supranormal physical qualities were frequently cited, such as the apparent invulnerability or indifference to the elements. Examples abound. One holy fool reportedly emerged unscathed after spending an entire night in a hot oven, and dirt never adhered to his feet even though he always walked barefoot. An Egyptian enraptured man sat in the sand all year long, another was said to sweat through the dead of winter despite wearing only a thin wrap, and yet another was reported to have spent five years of his life in an orchard, never leaving it, even when his body became covered in snow.[44]

Some holy fools were credited with paranormal and psychic abilities like clairvoyance, as they could predict future events or know the otherwise unknowable. 'Abd al-Ghani al-Nabulusi (d. 1731) recounted that as he was approaching the town of Jinin in Palestine, a group of holy fools came out to welcome him despite the fact that he had sent no word of his impending arrival. This otherwise inexplicable foreknowledge, he remarked, was proof of their powers.[45] The Egyptian

44 Al-Ghazzi, *Lutf al-Samar*, 1:306; al-Sha'rani, *al-Tabaqat al-Kubra*, 2:252; al-Muradi, *Silk al-Durar*; 3:244; and al-Burini, *Tarajim al-A'yan*, 2:163–164, respectively. In addition, the Aleppine Abu Bakr b. Abi al-Wafa' (d. 1583) was known to sleep on the bare ground under the open sky, inured to the hardships of heat, cold, wind, or snow. Further proof of his holiness was that in spite of the infliction of numerous sword wounds during a violent assault, he made a complete recovery and his scars miraculously cleared. In al-'Urdi, *Ma'adin al-Dhahab*, 46.

45 Al-Nabulusi, *al-Hadra al-Unsiyya*, 61. Elizabeth Sirriyeh points out that al-Nabulusi reported encountering an unusually high number of *majādhīb* in Palestine, certainly far more than during his travels in Lebanon, Egypt, and the Hijaz. Palestine's apparently large population of holy fools merits further investigation. See Elizabeth Sirriyeh, *Sufi Visionary of Ottoman Damascus: 'Abd al-Ghanī al-Nābulusī, 1641–1731* (New York: RoutledgeCurzon, 2004), 115. Another example of a holy fool's clairvoyance is in al-Muhibbi, *Khulasat al-Athar*, 1:374.

Ibrahim the Naked (d. 1520s) knew the full names of everyone he met without ever having been introduced to them, and in seventeenth-century Cairo, Isma'il al-Sanafiri foretold the imminent downfall of a particular Ottoman governor and accurately predicted the name of his replacement.[46] The Syrian al-Kiki (d. 1749) one day began to scream and sob on sighting a yogurt seller in the street, then demanded a pot of yogurt. Someone from the crowd now gathered around him complied, but al-Kiki rejected this offering and continued to wail, pointing to a specific pot and insisting that he be given it. On receiving it the majdhub emptied the pot's contents onto the ground and a large snake came slithering out of the yogurt. This exploit was interpreted as evidence of al-Kiki's ability to see things hidden from others, and thus of his powerful divine attraction.[47]

At times, wondrous deeds were truly a spectacle. The Cairene Ahmad Abu Shusha (d. 1703) was apparently able to talk, eat, and drink with one hundred needles stuck in his mouth.[48] Holy fools belonging to the Taghlibi clan in eighteenth-century Damascus and Cairo claimed powers that seemed nothing short of manipulating nature itself, including a ceremony called *dawsa*, "trampling," that consisted in riding horses over the backs of people lying on the ground without harming them. They also ate snakes, walked through fire, and could open securely locked doors with a tap of the head. The Taghlibis were further reputed to possess healing powers, among them, interestingly enough, the ability to restore clarity of mind to the violently insane.[49]

More often than not, however, divine attraction manifested itself in unconventional or idiosyncratic conduct that could take a multitude of shapes. It could be relatively benign and innocuous: remaining silent most of the time, marked incoherence of speech, speaking loudly in inappropriate settings, or screaming frequently for no apparent reason.

46 Al-Sha'rani, *al-Tabaqat al-Kubra*, 2:254; Çelebi, *Al-Rihla*, 2:84.

47 Shihab al-Din Ahmad Ibn Budayr al-Hallaq (fl. 1762), *Hawadith Dimashq al-Sham al-Yawmiyya min Sanat 1154 ila Sanat 1176*, MS, Chester Beatty Library, Dublin, Ar. 3551(2), 50b.

48 Al-Jabarti, *'Abd al-Rahman al-Jabarti's History of Egypt*, 1:114.

49 Muhammad Ibn Kannan al-Salihi (d. 1740), *al-Hawadith al-Yawmiyya min Tarikh Ihda 'Ashar wa-Alf wa-Mi'a*, partial edition by Akram Hasan al-'Ulabi with the title *Yawmiyyat Shamiyya* (Damascus: Dar al-Tabba', 1994), 118–119; al-Muradi, *Silk al-Durar*, 4:254–255.

Sulayman al-majdhub (d. 1773/4) was known for rapid and frantic patterns of speech. He was terrified of children and would scream and run away whenever they approached him. Among his oddities was to ask passersby for a specific small coin of little value. If instead given more valuable coins, Sulayman would chase his benefactor, return the coin in question, and explain that he only accepted that specific small coin.[50]

Another majdhub was in the habit of taking myrtle from gravestones, and Ahmad al-Sarkhadi roamed the streets of Damascus carrying a censer so as to perfume those he encountered with incense.[51] The Aleppine Khalil al-Kurdi (d. 1599/1600) was frequently seen in a local mosque where he would perform all five required daily prayers, but his body convulsed so "violently because of the severity of his divine attraction" that his prayers were invalidated.[52] The Egyptian ʿAbd al-Rahman al-majdhub (d. 1538) would speak for three months and then would be silent for three months, and referred to himself in the third person.[53] And a much-revered Damascene majdhub displayed his divine attraction through provocative and enigmatic statements. One of his memorable pronouncements began with the pedestrian declaration, "I want a woman to fuck" (*biddī mar'a ḥattā anīkahā*), an announcement that was accompanied by copious weeping. When asked to explain himself, he provided a cryptic coda, asserting simply "a woman is a loaf of bread."[54]

The peculiarities of holy fools, their frequently unconventional physical appearance and grooming, and their wondrous deeds invite

[50] Al-Muradi, *Silk al-Durar*, 2:208.

[51] In al-Ghazzı, *Lutf al-Samar*, 2:606 and 1:307, respectively.

[52] Al-ʿUrdi, *Maʿadin al-Dhahab*, 343.

[53] Al-Shaʿrani, *al-Tabaqat al-Kubra*, 2:252. For more examples of eccentric behavior, see al-Burini, *Tarajim al-Aʿyan*, 2:162; al-Muhibbi, *Khulasat al-Athar*, 4:511; al-Ghazzi, *al-Kawakib al-Sa'ira*, 2:72; al-ʿUrdi, *Maʿadin al-Dhahab*, 343. Especially interesting is the sixteenth-century scholar Mirza al-Shirazi, who was described as "not devoid of divine attraction." The condition was deemed responsible for his "astonishing thoughts" and his tendency to behave unconventionally. For instance, while acting as *naqīb al-ashrāf* (head of the descendants of the Prophet) at the Imperial Palace in Istanbul, al-Shirazi sent a message to the Sultan with the bold request that he be granted access to the Porte at any time of day or night, so that he might occasionally bring the Sultan some of the dishes he had prepared with his own hands. The Sultan complied with this extraordinary request, which surprised al-Shirazi's biographer only slightly less than the fact that the ruler actually deigned to eat the food prepared by al-Shirazi. In al-Burini, *Tarajim al-Aʿyan*, 2:54–55.

[54] Ibn Budayr al-Hallaq, *Hawadith Dimashq al-Sham*, 50b.

associations with ecstatic and antinomian Sufi individuals and groups who purposefully defied social conventions and often Islamic law in their quest to focus exclusively on achieving mystical union with God.

Some holy fools did in fact belong to the numerous unorthodox groups that, for reasons that await thorough investigation, expanded considerably in the Ottoman period. In addition to "classic" antinomian orders like the Qalandariyya (a mendicant order whose followers shaved their heads and their facial hair and wore tattered garments) and the Malamatiyya (the "Blame-Seekers," because they strove to attract blame and contempt from society in order to suppress their egos and thus facilitate the achievement of closeness to God), many other ecstatic and antinomian Sufi orders thrived in the Ottoman Arab world. The Rifa'iyya was associated with snake charmers and with the physical violence its adherents inflicted on themselves. The Sa'diyya, founded in Syria as a subbranch of the Rifa'iyya, was characterized by the horse "trampling" ceremony mentioned earlier. The Ahmadiyya was especially popular in Egypt, where it appealed to the lower classes of society and attracted the scorn and ire of the educated elites.[55]

The obvious similarities between several holy fools described in this chapter and the followers of ecstatic and antinomian Sufi orders, however, should not tempt us to conflate the two categories wholesale. One fundamental difference is that a majdhub was *involuntarily* attracted to God and did not necessarily choose his way of life. This difference must have been obvious to early modern Muslims, for there are plenty of examples of men who are identified by their biographers as affiliated with an antinomian or other Sufi order but who are not labeled holy fools. Conversely, many, if not most, holy fools were not associated with a Sufi order.[56] In other words, a person could be an ecstatic or antinomian Sufi but not a majdhub, and vice versa.

This distinction bears emphasizing because conflating the two categories obscures a crucial point: "majdhub" was an umbrella category

[55] Michael Winter, *Egyptian Society Under Ottoman Rule, 1517–1798* (London and New York: Routledge, 1992), 128–166. See also Ahmet Karamustafa, *God's Unruly Friends: Dervish Groups in the Islamic Later Middle Period, 1200–1550* (Salt Lake City: University of Utah Press, 1994).

[56] Evliya Çelebi, for instance, clearly differentiated between the Malamatis and the holy fools of seventeenth-century Cairo and chose to dedicate two separate sections of his book to them. Çelebi, *al-Rihla*, 2:83–84.

that encompassed mental disturbances and impairments that could not easily be labeled as depressive melancholic ailments or as violent madness.

Women and Holy Folly

Holy fools hailed from all social and economic classes. Some came from privileged backgrounds and were respected scholars, teachers, or local notables when they were overcome by divine attraction. A handful held high office in local, provincial, or imperial government. Some were judges.[57] But the majority can hardly be described as the product of the upper classes. We read of grocers, water carriers, tailors, weavers, butter vendors, slaves, of a man who earned his living by pressing oil from sesame seeds, of a soldier who worked as a courthouse bailiff, and even of a bandit in the Egyptian Nile Delta region.[58]

Although they came from all walks of life, holy fools were overwhelmingly men. Such a high level of gender disparity is proportional to the general sparseness of references to women in early modern Arabic sources, but it is also doubtlessly a function of the fact that it was far more difficult for women to transgress the confines of propriety dictated by socially constructed gender roles. For instance, a female holy fool found wandering in the streets naked would have been much more likely than her male counterpart to be declared insane and hence to be confined to her relatives' home or to a hospital. It is then no coincidence that the few female holy fools mentioned in Arab biographical dictionaries, chronicles, and travelogues are usually the wives of holy fools or are otherwise associated with one.[59]

The spouses of holy fools were allowed wider latitude of behavior than other women, provided they had the support of their husbands. A fascinating example is the former wife of a prominent judge in

[57] Al-Burini, *Tarajim al-A'yan*, 2:52–56; al-Ghazzi, *al-Kawakib al-Sa'ira*, 2:87; al-Ghazzi, *Lutf al-Samar*, 1:248; al-'Urdi, *Ma'adin al-Dhahab*, 66, 259.

[58] Al-'Urdi, *Ma'adin al-Dhahab*, 45, 178, 343; al-Ghazzi, *Lutf al-Samar*, 1:258, 307; al-Ghazzi, *al-Kawakib al-Sa'ira*, 2:250; al-Muradi, *Silk al-Durar*, 2:93; al-Nabulusi, *al-Hadra al-Unsiyya*, 67, 69; al-Muhibbi, *Khulasat al-Athar*, 2:111.

[59] See, for example, the wife of a majdhub in seventeenth-century 'Ajlun, in today's Jordan. Her husband's name was Muhammad b. Humud. It appears that Muhammad b. Humud came from a family of holy fools that included enraptured brothers and cousins. Al-Nabulusi, *al-Hadra al-Unsiyya*, 63.

Damascus. We are informed that she, whose name is never stated, was a "secluded woman" (*mukhaddara*). She was fully veiled, rarely left her home, and belonged to the highest stratum of society. Somehow she made the acquaintance of a highly revered majdhub known as al-Zughbi (d. 1570), who is described as "one of the marvels of Damascus." Apparently the lady concurred with the description, for she proposed marriage. He accepted, and the marriage was pronounced one of his "wonders," perhaps because he married in the first place (many holy fools never wed), because he married so far above his station, or because of the unconventional nature of the marital relationship he formed with his bride. The lady was overcome by divine attraction shortly after her marriage and she adopted al-Zughbi's mode of life. She liquidated her substantial properties, abandoned her lavish lifestyle and her fancy clothes and jewelry, ceased to wear the face veil, and accompanied her husband wherever he ventured, entering mosques and private homes with her face uncovered. The couple provoked additional amazement by often sharing a bed and failing to use a privacy screen when they were overnight guests at a preacher's home. Their conduct as a couple and her comportment as a woman were shocking to many. Al-Zughbi was assailed for permitting his wife to flout the conventions of her class and gender so brazenly, but he was unmoved by such criticism, telling his followers that they should regard her as their spiritual mother. He was so devoted to his spouse that when she died he displayed great sadness and passed away himself a year later.[60]

It was not impossible for women to attain the status of holy fools in their own right, but in order to gain the stamp of legitimacy they might need to first be closely connected to a male holy fool. The Egyptian historian al-Jabarti reports that in the spring of 1786 there took place in Cairo what was in his estimation "a disgraceful event." A woman called Shaykha Ammuna became attached to the popular majdhub Shaykh ʿAli al-Bakri (there is no intimation that the relationship was of a romantic or sexual nature). Fully veiled at first, she followed him everywhere, including into people's homes, where she had easy access to the women's quarters. Taking her muddled speech as a sign of

[60] There are two biographies of al-Zughbi, each providing different details. Al-Burini, *Tarajim al-Aʿyan*, 3:162b–163b, and al-Ghazzi, *al-Kawakib al-Saʾira*, 3:32–34, 3:80.

saintliness, al-Jabarti continues, the women she met started to believe in her and would give her money and clothes. The news spread that Shaykh 'Ali had transmitted divine attraction to her through a glance. As God's attraction increased, she cast off the face veil and started to accompany Shaykh 'Ali "dressed like a man." The followers of the enraptured pair steadily grew to the point that wherever they went crowds gathered around them, causing commotion. During these gatherings Shaykha Ammuna would stand up on a bench and "speak indecently," at times in Arabic and at times in Turkish. We are not informed of what she said, but it must have impressed many of her listeners, who rushed to kiss her hands in hopes that her blessing might be transmitted to them.

But not everyone was enthusiastic. One day Shaykha Ammuna, Shaykh 'Ali, and their entourage were seized by a soldier while they were passing through the lane in which he lived. Shaykh 'Ali was fed and let go, but Shaykha Ammuna and the other holy fools accompanying her were imprisoned and beaten. After they asked for forgiveness, repented, and put on clothes, the soldier set the other holy fools free but sent Shaykha Ammuna to a hospital, where he had her shackled with the insane for some time. She was later allowed to leave the hospital in the wake of "certain events" that are not disclosed, and eventually became an object of veneration for many men and women independently from Shaykh 'Ali.[61] Although she may be regarded as a success story, the price she had to pay because she was a woman who overstepped the boundaries of her gender role was high: imprisonment, beatings, and hospital confinement in chains.

Impostors?

Despite transgressions of cultural norms, bizarre and potentially destabilizing public behavior, and the performance of acts that verged on sorcery, the veneration of holy fools cut across lines of class and was by no means restricted to the lower orders of society. Regardless of educational level or socioeconomic status, everyone believed in holy fools as a category. All the Ottoman Arab chroniclers, biographers,

[61] Al-Jabarti, *'Aja'ib al-Athar*, 3:322–323, 4:237–238. I am loosely paraphrasing the story.

and travel writers I have examined deemed them worthy of inclusion in their works.

Acceptance of holy folly as an authentic manifestation of the divine also cut across lines of gender. Some appealed especially to women. Hasan al-majdhub (d. 1609), from the Palestinian village of Zayta, near Nablus, became the object of female adoration after his arrival in Damascus. He was reportedly visited by flocks of gift-bearing women with whom he publicly exchanged kisses, a practice that raised several eyebrows.[62]

This does not mean, however, that all holy fools were uncritically accepted at face value. Remarks such as "people believed in him" customarily accompany discussions of holy fools and are often tempered by qualifiers like "some," "many," "most," or enhanced by phrases such as "there was no question about his saintliness," indicating that the majdhub's bona fides were vulnerable to suspicions of invalidity or even fraud.[63] Some members of the educated elite considered themselves less prone than the general populace to uncritical acceptance of the saintliness and piety of specific holy fools. The mufti and biographer Muhammad Khalil al-Muradi (d. 1791) was wont to specify that belief in a particular majdhub was shared by the general public

[62] Al-Burini, *Tarajim al-A'yan*, 2:163. Abu Bakr b. Abi al-Wafa' (d. 1583), Aleppo's most famous holy fool during the four centuries of Ottoman rule of that city, also attracted a very large following of women. His lodge on the northern outskirts of Aleppo was still operating in the middle of the eighteenth century, when it was "visited by vast crowds of women, two or three times in the year." Russell, *The Natural History of Aleppo*, 1:257. See also ibid., 1:207. Indeed, a cult emerged around him that lasted well after his death in 1583. William Biddulph, who served as a chaplain to the English merchants in Aleppo in the early 1600s, provided a (rather contemptuous) description of the veneration accorded to the followers of this holy fool. William Biddulph, "Part of a Letter of Master William Biddulph from Aleppo," in Samuel Purchas, *Hakluytus Posthumus, or, Purchas His Pilgrimes: Contayning a History of the World in Sea Voyages and Lande Travells by Englishmen and Others* (Glasgow: James MacLehose and Sons, 1905), 8:263–264. Heghnar Watenpaugh states that Abu Bakr's lodge continued to be a place of veneration into the twentieth century. See her "Deviant Dervishes: Space, Gender, and the Construction of Antinomian Piety in Ottoman Aleppo," *International Journal of Middle East Studies* 37, no. 4 (2005): 536.

[63] These suspicions could have fatal consequences. In 1523, a holy fool named Dhu al-Nun al-Kalamani was murdered by peasants in the mountains near Damascus. The incident was reported by Ahmad b. Muhammad Ibn al-Himsi (d. 1527/8) in *Hawadith al-Zaman wa-Wafayat al-Shuyukh wa-l-Aqran*, ed. 'Umar 'Abd al-Salam al-Tadmuri (Sidon; Beirut: al-Maktaba al-'Asriyya, 1999), 3:53–54, and repeated by al-Ghazzi in *al-Kawakib al-Sa'ira*, 1:194.

(*al-ʿāmma*) as well as the elite (*al-khāṣṣa*). He thus attested to the given individual's credentials by observing that belief in his divine attraction was not restricted to the more ignorant and superstitious segments of the population.[64]

For reasons that remain to be investigated, Egyptian writers tended to voice more skepticism of holy fools than their Levantine peers. In his biographical dictionary, ʿAbd al-Wahhab al-Shaʿrani (d. 1565) was careful to discriminate between genuine holy fools and the fraudulent ones, whom he abhorred. Two centuries later, the Egyptian historian al-Jabarti, too, was openly critical of certain holy fools, including the very popular Shaykh ʿAli al-Bakri (d. 1792/3) described earlier. Al-Jabarti referred to him as a "halfwit," adding contemptuously that "people had great faith in him, they would listen to his ravings, directing his expressions, and explaining them according to their own objectives and needs and occurrences."[65] Most vitriolic of all was the prolific and generally caustic Cairo-based poet Hasan al-Badri al-Hijazi (d. 1718/9), who is frequently quoted by al-Jabarti admiringly. Al-Hijazi did not mince his words: "Would that we had not lived to see every madman become a saint in the eyes of the people."[66]

Overall, however, outright attacks on individual holy fools were rare because of the risk of attracting a majdhub's vengefulness. This happened to none other than Najm al-Din al-Ghazzi, our oft-cited biographer. Al-Ghazzi was regarded by his peers as the foremost Hadith scholar in Damascus. One day, as he was in the middle of holding a Hadith lesson in the Umayyad Mosque, a majdhub nicknamed Ibn Farfara (d. 1656/7) began to pose seemingly irrelevant questions and utter strange words. Losing his patience, al-Ghazzi bluntly told him to shut up. Ibn Farfara immediately retorted angrily "No, you shut up!" and stormed out of the mosque. A few days later Najm al-Din al-Ghazzi was struck by a mild form of hemiplegia (partial paralysis) that caused his speech to become impaired. Many interpreted Najm al-Din's ailment as God's way of punishing him by, quite literally, shutting him up for mistreating a holy fool. This interpretation was corroborated by the fact that Najm al-Din did not regain the ability to

64 Al-Muradi, *Silk al-Durar*, 1:50, 2:93, 4:255.
65 Al-Jabarti, *ʿAbd al-Rahman al-Jabarti's History of Egypt*, 3:126, 2:412.
66 Al-Jabarti, *ʿAjaʾib al-Athar*, 5:86.

speak fluently until he reconciled with Ibn Farfara several years later. Subsequently, it was said, al-Ghazzi often apologized to him and kissed his hand, forever considering him dear to his heart.[67]

What accounted for the tolerance of holy folly even among skeptics was precisely the popular belief that because of the intimate bond between holy fools and the divine, all who abuse, oppress, or otherwise mistreat a majdhub will suffer dire consequences, regardless of their rank, wealth, or power. One final illustrative example is that of Abu Bakr b. Abi al-Wafa' of Aleppo (d. 1583). Perhaps the quintessential holy fool, he embodied the most extreme forms of divine attraction. A bulky man who never married, always remained clean-shaven, slept outdoors throughout the year, declined to pray or fast, and was known for violent mood swings, Abu Bakr took it to the next level, as it were, by piercing his ears, reportedly pulling all of his own teeth in a single day, spending much of his life surrounded by and caring for a pack of dogs, and routinely addressing men in the feminine.

Reports of his behavior caused state officials and other notables to suspect that Abu Bakr was simply insane and ought to be confined to a hospital. Their suspicions were usually allayed when they met him, and most people continued to believe in his saintliness. But on at least one occasion Abu Bakr narrowly escaped execution. Hasan Pasha, the son of the powerful governor of Aleppo, was curious and sought out the holy fool. On introduction to Hasan Pasha, described as a comely beardless youth, Abu Bakr unceremoniously fondled the young man's buttocks and made a rude, vulgar comment to the effect of "this is what you are good for." Enraged by the affront, Hasan Pasha wanted to have Abu Bakr killed. He only desisted because his companions convinced him that taking his life would provoke God's wrath and would inevitably result in his executioner's destruction.[68]

[67] Al-Muhibbi, *Khulasat al-Athar*, 2:112, 4:199–200. The same moral is conveyed by another story, albeit one that does not close with a happy ending. Abu al-Wafa' al-'Urdi (d. 1660) was an eyewitness when a drunken man stabbed the majdhub Khalil al-Kurdi (d. 1599/1600) in Aleppo. While al-Kurdi survived the assault, his attacker perished shortly thereafter, an outcome that was numbered among al-Kurdi's miracles. Al-'Urdi also reports that the attacker's "clan soon lost power." *Ma'adin al-Dhahab*, 343. For another example of the dangerous consequences of abusing a holy fool, see al-Nabulusi, *al-Hadra al-Unsiyya*, 69.

[68] The several extant biographical accounts of Abu Bakr b. Abi al-Wafa' provide complementary and contradictory details. I have relied on al-'Urdi, *Ma'adin al-Dhahab*,

The denouement of this tale identifies the issue most critical to popular and official responses to the unconventional and often outlandish behavior of holy fools: how could anyone, regardless of social status or power, be absolutely sure that a majdhub "in whom people believed" was *not* a true friend of God? Why risk offending the general populace, an outcome with inherent potential for social destabilization, or, perhaps even more significantly, incur the displeasure of God with possibly disastrous consequences in this world and the next? The risks were so great that drastic measures against holy fools for offenses at the level of a personal slight, however egregious and outrageous, were rarely taken.

The Law and the Limits of State Tolerance

Holy fools occupied a special legal zone that they shared with all others deemed to be deprived of reason. Just as impairments of the mind were ubiquitous in Ottoman Arab homes and streets, they were also a central concern of Muslim jurists and were ubiquitous in Islamic law. In addition to the extensive treatment they received in law books, Ottoman-era fatwa collections contain copious questions and answers pertaining to mental disturbances.

There were points of disagreement among the four main Sunni schools of law, but the overarching principle on which they all concurred was that a person bereft of the capacity to reason had no individual responsibility before the law. As far as the law was concerned, the causes and types of mental incapacitation were irrelevant. The same rules applied whether it was due to an epileptic fit, idiocy (*'atāha*), madness (*junūn*), divine attraction (*jadhb*), stupor (*dhuhūl*), or other factors.

Once the "departure of reason" (*zawāl al-'aql*) was determined, the individual's legal status became that of a minor child. Persons so judged were under no obligation to pray or to perform any other ritual observance of the faith, and they were barred from testimony in court. A legal guardian was appointed to manage their financial affairs so that, at least in theory, their money and property would not be lost.

43–54; and 241–252 of the Abu Salim edition (see bibliography); al-Ghazzi, *al-Kawakib al-Sa'ira*, 3:98–99; Ibn al-Hanbali, *Durr al-Habab*, 1:394.

Their closest relatives were responsible for financially supporting them if needed. Just like minor children, their legal guardians could also marry them off without their consent.

Over and over again, muftis ruled that holy fools, idiots, and madmen were legally minors. This status, however, was not irrevocable, and in periods of lucidity a person could regain his or her legal rights and responsibilities. The "departure of the mind" might be a permanent condition or it might be intermittent, and, depending on its etiology, might even disappear like any other illness.[69]

Fatwa collections are replete with questions about mental incapacity, especially in the context of marriage and divorce. Many of the queries addressed to muftis concerned the validity of divorce pronouncements uttered by a husband during an epileptic fit, a state of lunacy, holy folly, mental confusion, an inflammation of the brain, or by a man deemed an imbecile. Invariably, the muftis replied that pronouncements made by men lacking reason were invalid. The only exception was the loss of reason as a result of drunkenness, for which there was no excuse.[70]

The exemption from individual responsibility extended to the sphere of criminal law and included homicide (unless committed during a period of lucidity). The fundamental principle was that without reason there can be no deliberate intent and thus no responsibility. For example, one day a Sufi killed a cat in the Yalbagha Mosque in Damascus, of which a certain Hasan al-majdhub was a habitué. When the man lay down to take a nap, Hasan smashed a rock on his head, killing him. Questioned by officials, Hasan acknowledged the act, which he justified by the simple statement that the Sufi "had killed his cat."

[69] A comprehensive treatment of insanity in medieval Islamic law can be found in Dols, *Majnūn*, 434–455. Mental disabilities in Islamic law are also discussed in Rispler-Chaim, *Disability in Islamic Law*.

[70] For a small sample of fatwas dealing with lack of reason in all its forms, see al-Tumurtashi, *Fatawa*, 68b, 108a–108b, 109a, 128a, 140b–141a; al-Tarabulusi, *Fatawa*, 58, 59, 66, 114, 115; Ibn al-Ha'ik, *Fatawa*, 4a, 4a–4b, 11a, 15a, 15b, 17a, 72b, 76b, 155b, 156a; al-Ramli, *al-Fatawa al-Khayriyya*, 1:34–35, 36–37, 38–39, 44, 68, 179; al-'Imadi (as edited by Ibn 'Abidin), *al-'Uqud al-Durriyya*, 1:38, 42, 45, 77, 81, 100, 258 and 2:51, 146, 146–147, 151, 257, 290, 304; al-Hanuti, *Muntakhab al-Fatawa*, 78b, 87b–88a, 108a; al-Husayni, *Fatawa*, 150a, 264a-b, 268a. For a discussion of the idiotic or feeble-minded, see Ibn 'Abidin, *Radd al-Muhtar*, 3:173–174.

He was declared to lack reason and thus unanswerable for his crime. Consequently, the authorities set him free.[71]

Holy fools enjoyed remarkable latitude in speech and behaved in ways that would customarily result in social sanction and exclusion, if not legal action. They could pray in a state of impurity (without having performed the necessary ablutions), neglect to pray altogether, disdain to fast during the holy month of Ramadan, and eat dog meat or the raw heart, liver, and lungs of animals, which were considered disgusting and impure.[72] And although violent behavior was far from the norm for most, it would be idealistic to ascribe a doctrine of nonviolence or passivity to holy fools.[73] Indeed, we have distressing reports of sexual assaults and rapes by enraptured men throughout the Ottoman period. An account from sixteenth-century Cairo involves a young woman who was sexually assaulted by a majdhub in the middle of the street as she was leaving a bathhouse. Shortly after the assault passersby flocked to her, wanting to touch her clothes that they believed blessed by contact with the majdhub. When apprised of the incident, her husband apparently rejoiced and took it as a good omen (no information is provided about how the woman herself reacted). The majdhub suffered no consequences.[74]

On occasion, as in the case of the majdhub who killed a Sufi who had killed a cat, holy fools could literally get away with murder. But there were limits to state tolerance. When Abu Bakr al-Armanazi (d. 1609/10), a notable from the small northwestern Syrian town of Armanaz, first manifested divine attraction, he shaved his beard. It was said that he intended to imitate a famous holy fool from Aleppo to whom he bore a physical resemblance. The biographical sketch that has survived to us is unflattering: al-Armanazi is described as a man who did not pray or fast and spoke and acted like "abominable heretics."

This indictment was produced by the cumulative effect of several incidents, each of which displayed al-Armanazi's disregard for the

[71] In al-Burini, *Tarajim al-A'yan*, 2:162–163, and in al-Ghazzi, *Lutf al-Samar*, 1:402–405.

[72] See, for instance, al-'Urdi, *Ma'adin al-Dhahab*, 178, 179.

[73] Dols, *Majnūn*, 422: "The holy fool was essentially harmless, physically if not psychologically."

[74] Dols, *Majnūn*, 413–415.

doctrinal and judicial authority of the educated elites. One incident featured the majdhub investigating fundamental issues of the faith that Muslim jurists and theologians had long settled. He reportedly communicated with a group of Jews, expressing his desire to "have a discussion with you and the Muslims, so that I can discern which party is right." (Al-Armanazi, incidentally, denied this particular allegation.) Such a proposition was heretical and blasphemous, for it questioned the superiority and perfection of Islam and presumed the majdhub's possession of spiritual knowledge and discernment superior to that of generations of Muslim scholars.

In another incident, al-Armanazi assaulted and injured the judge of Harim, a small town near Armanaz, during a long period of drought. He justified his action by asserting that rain would not come to the area until the judge in question was struck. The judge of Harim was not pleased, unsurprisingly, and lodged a complaint against al-Armanazi with the far more powerful judge of Aleppo. Since he was a friend of al-Armanazi's brother-in-law, the judge of Aleppo sought to prevent the holy fool's destruction in the following manner.[75] He advised his friend to coach al-Armanazi and instruct him to jabber senselessly in court, thus providing a pretext for declaring him insane (*majnūn*). Significantly, it was judges who were empowered to decide the fate of supposedly mentally incapacitated people who came to their attention; physicians were not routinely consulted to determine if an individual was insane. The judge would then be able to order his confinement in a hospital instead of sentencing him to prison or even sending him to the gallows.

Al-Armanazi refused to comply with the judge's instructions, but the judge nevertheless found a way to justify declaring him insane and ordered his temporary confinement in an asylum. The judge's leniency, however, merely delayed al-Armanazi's downfall. The stage for his demise was set when a new governor was appointed to the province of Aleppo, altering the delicate balance of power in the region. When another Ottoman official sought control over the municipal revenues

[75] Al-Armanazi's relative is described as his *ṣihr*, which can mean either brother-in-law or son-in-law. Given the context, I have opted for what seems to me the more likely of the two possibilities.

of the town of al-Armanaz, the majdhub organized opposition. In an unrelated dispute, al-Armanazi had in the meantime alienated his influential brother-in-law. Having now made powerful enemies, both old and new, and having lost his protector, al-Armanazi was doomed. The deputy judge of Aleppo, who happened to be a brother of the judge of Harim who had been assaulted by al-Armanazi, various other Ottoman officials, and village notables now gathered witnesses willing to give evidence of the holy fool's heretical words and scandalous deeds. The court heard testimony that al-Armanazi drank wine, refused to pray or fast, uttered blasphemous declarations like "I am a Prophet," and openly questioned the spiritual and legal superiority of Muslims over Jews.

The collective weight of the evidence and the power and hostility of those sitting in judgment spelled the end of this majdhub. When brought to the place of execution under the Citadel of Aleppo, al-Armanazi was heard reciting the Qur'an, seemingly unperturbed. His head was then struck from his body, which was daubed with tar and set ablaze. According to his biographer, "most people" believed that al-Armanazi deserved his fate, but a minority saw him as the victim of injustice.[76]

Publicly declaring oneself to be a prophet was an extraordinarily provocative act, as was questioning the doctrinal superiority of Islam or the legally sanctioned privileged status of its believers. In the Ottoman Arab world, like elsewhere, these were red lines that simply could not be crossed, regardless of wealth, social standing, or even holiness. Exacerbating these offenses were the powerful temporal forces arrayed against al-Armanazi (largely of his own making). For these reasons Abu Bakr al-Armanazi was executed. Having gone too far and having lost the protective status of majdhub, his only remaining option was to pretend to be insane and be taken to a hospital and thus save

[76] Al-Armanazi's biography appears in al-'Urdi's *Ma'adin al-Dhahab*, 65–68. I have compared Altunji's edition with Abu Salim's and al-Ghazali's editions (see bibliography), 259–261, and 15–18, respectively. Each version provides slightly different but complementary details of the story. See also the story of Sulayman, a holy fool who, in August 1807, attracted a large following in the northeastern Egyptian countryside, creating unrest. His tale did not have a happy ending, either. Al-Jabarti, *'Abd al-Rahman al-Jabarti's History of Egypt*, 4:90–93.

his own life. He refused to do so, and in the end he was dealt with not as a holy fool, not as a madman, but as a heretic.[77]

Conclusion

Besides idiocy, usually regarded as a congenital and permanent condition, impairments of the mind were grouped into three broad and sometimes overlapping categories: melancholia, lunacy or violent madness, and holy folly. Melancholia was an expansive and multifaceted concept that encompassed a broad spectrum of symptoms and behaviors that were not normally associated with sin, guilt, divine punishment, demons, or other supernatural forces. Melancholia was categorically described as an illness rooted in physiological processes easily explicable within the framework of humoral medicine, and as such it was not customarily shrouded in secrecy or shame. The result of a surfeit or putrefaction of "dry" bile, typically black but occasionally yellow, it could be caused by a variety of environmental and behavioral

[77] The life of Nizam al-Din al-Sindi (d. ca. 1607) also exemplifies the dangerous porousness of the boundaries between madness and heresy. Al-Sindi belonged to the Naqshabandi Sufi order and, as his name indicates, he began his life in al-Sind, an area that roughly corresponds to the southern region of Pakistan. His generally hostile biographers state that the "clever" al-Sindi arrived in Damascus with his younger brother and proceeded to do something that was declared astonishing: he proclaimed himself the Mahdi, the "divinely guided one" or Messiah. When told that he failed to meet the criteria for this role, Nizam al-Din had ready answers. Someone observed that, contrary to tradition, his name was not Muhammad. Al-Sindi replied that "Nizam al-Din" was an honorific title of the Prophet. Others pointed out that he could not possibly be a descendant of the Prophet Muhammad, since he was "a black Sindi." In response, he asserted that he was indeed a direct descendant of the Prophet but had left behind the documents proving the claim, and that it was not the color of his skin that mattered, but rather "the metaphorical whiteness that is manifested in [one's] deeds." Despite these skeptics, al-Sindi enjoyed popularity and, emboldened by his followers' support, proclaimed from a minaret of the Umayyad Mosque: "Oh people of Damascus, I am the Mahdi of Time, and I summon you to answer my call and follow me!" Decreed to have gone mad (*janna*), the Chief Judge of Damascus confined him to an asylum, permitting his release only after he recovered from his mental confusion (*takhlīṭ*). After his release, al-Sindi fled Damascus and traveled to Jerusalem, Nablus, and Gaza, where again he fought with local scholars, and finally to Cairo, where he died shortly after his arrival. His biography is recorded in al-Burini, *Tarajim al-A'yan*, 3:174a–174b, and al-Muhibbi, *Khulasat al-Athar*, 4:451–453. Al-Muhibbi drew most of his information from al-Burini but provided supplementary details including al-Sindi's scholarly credentials, and was less openly hostile to him.

factors. Those suffering from the mildest forms of melancholia exhibited symptoms nowadays associated with depression, and the more severe forms could produce delusional thoughts, obsessive fixations, compulsive behavior, or seemingly senseless violence inflicted on the self or on others.

The most violent forms of melancholia tended to be conflated with lunacy (*junūn*), a condition that proves resistant to precise definition. Despite the shared semantic derivation of the terms *junūn* and jinn, invisible spirits are rarely invoked as progenitors of madness. Instead, the ubiquitous humoral imbalance was more commonly cited. The paucity of persons labeled lunatics in early modern Arabic sources, especially compared to the abundance of holy fools, may stem in part from a general reluctance to affix a label so potentially disabling, as it was fraught with grievous legal and social consequences.

Lastly, holy folly was clearly deemed a form of madness but was not conceived as an illness requiring or amenable to any form of medical treatment. On the contrary, the condition's divine origin elicited a measure of reverence from all strata of society. Acquiring the status of majdhub was by no means a transparent or uncontested process. A general consensus about the authenticity of the individual's saintliness and wondrous deeds was necessary before such a designation could be made. Individual holy fools might be liable to accusations that they were (1) charlatans who were merely simulating divine attraction; (2) deranged persons whose madness was manifested in provocative words and deeds that could be misconstrued as evidence of divine attraction; or (3) heretics whose unorthodox beliefs and practices bore superficial similarities to those of genuine holy fools. Evidently, the boundaries demarcating the holy fool from the insane and the heretic were in some cases negotiable. As the rise and fall of Abu Bakr al-Armanazi shows most dramatically, this conceptual fluidity could have disastrous consequences. Ultimately, the holy fool was an umbrella category, a conceptual vessel capable of accommodating a host of unconventional and even violent behaviors that would have provoked serious social, if not legal, sanctions had they not been afforded this sanctifying imprimatur. Thus could the disruptive potential of numerous eccentricities be contained and social stability maintained.

4

Intersex

In the year 1545 or 1546, a brown-skinned, beardless youth called ʿAli worked as a bookbinder in the Qaymariyya quarter of Damascus. A man by the name of ʿAbd al-Rahman was in love with him. The matter came to the attention of the Shafiʿi deputy judge of the nearby neighborhood of Midan, who asked ʿAli to appear in court. On seeing him, the judge developed the impression that ʿAli "tended toward the feminine" and suggested that he was probably a *khunthā* (intersex). To remove any doubt, the judge tasked some physicians with conducting a medical examination of the boy. The doctors discovered that ʿAli had a vulva covered by "a small nipple" on which were three small holes. They proceeded to excise the extra tissue, and beneath it there indeed appeared a female vulva. Thereupon, the judge ruled that ʿAli was actually a girl. His name was changed to ʿAliyya and she was married off to ʿAbd al-Rahman, the man so besotted with him/her. When they consummated the marriage, ʿAbd al-Rahman "removed her virginity." ʿAliyya became pregnant and bore her husband the first of several children. "And this was witnessed and verified by most of the people of Damascus."[1]

While today deafness, blindess, and impairments of the mind are usually thought of as disabilities, intersex rarely is. Yet in early modern Arab societies there were spheres of life in which people with

[1] This story, which I am paraphrasing, was narrated by Muhammad b. Muflih al-Ramini (1524–1603), a historian and Hanbali judge from Damascus, and reported by al-Muhibbi in *Khulasat al-Athar*, 3:316. The full name of the boy is ʿAli b. al-Rifaʿi, of the man ʿAbd al-Rahman b. al-Zanni (or al-Zinni), and of the judge Kamal al-Din al-ʿAdawi al-Biqaʿi.

atypical or ambiguous genitalia were much more impaired than people who were deaf, blind, or mad. Hence, there is no reason to *not* regard intersex conditions as impairments. This chapter is dedicated to exploring how Arab Muslims understood, described, treated, and legislated sex indeterminacy.

Disorders of Sex Development in the Contemporary Arab World

To make sense of the vignette of ʿAli/ʿAliyya above, it may be useful to begin with a brief overview of intersex conditions, also known today as disorders of sex development (DSD), in the Arab world today.[2] This book is not a medical history of disabilities, and it is not my intention to present a medicalized account of intersex. However, only if we are aware of the link between consanguineous marriages and DSD can we fully appreciate why Muslim jurists, as well as other writers, paid considerable attention to ambiguous genitalia and dedicated special chapters of their works to them, leaving modern Western scholars confounded.[3]

[2] Until less than 100 years ago intersex conditions were known as hermaphroditism. In the United States, starting in the 1950s the word "hermaphrodite" came under sustained attack by activists on the grounds that it was not only stigmatizing but also descriptively inaccurate, for it suggests that there are people who have both female and male sets of genitals, an impossibility in humans. The term "intersex" was adopted in its place because it was found to be more precise. Within a few decades, however, the term "intersex," too, was attacked for being only marginally less stigmatizing than "hermaphrodite," and in 2006 a new, clinically based coinage, disorders of sex development (DSD), emerged. But this new nomenclature has by no means been accepted by all. The main objection is that "disorder" implies that the condition requires medical attention and correction, which is not always the case. Elizabeth Reis proposes instead "a new term, still DSD, but standing for divergence of sex development. By using this nomenclature, doctors, patients, and parents can articulate difference but not disorder." Elizabeth Reis, *Bodies in Doubt: An American History of Intersex* (Baltimore: The Johns Hopkins University Press, 2009), xv. As of the time of writing, there is little consensus beyond the undesirability of "hermaphrodite," and both "intersex" and "disorders of sex development" are used.

[3] See Charles Hamilton's comment that "*Hermaphrodites* are probably a class of beings which exist in imagination rather than in reality" in ʿAli b. Abi Bakr al-Marghinani, *The Hedàya, or Guide; A Commentary on the Mussulman Laws: Translated by the Order of the Governor-General and Council of Bengal by Charles Hamilton* (London: Printed by T. Bensley, 1791), 1:lxxv. Italics in the original. See also Colin Imber, *Ebu's-su'ud: The Islamic Legal Tradition* (Stanford, CA: Stanford University Press, 1997),

The clinical data currently available shows that the prevalence of urogenital anomalies and disorders of sex development (DSD) in Arab countries is considerably higher than in much of the rest of the world. Several neonatal and infant screening programs were conducted between 2000 and 2009 in clinics and hospitals in Egypt – Greater Cairo, Alexandria, and the Qalyubiyya governorate in the Nile Delta – with the purpose of determining the prevalence of DSD. All the studies concluded that the rate of urogenital anomalies in Egypt is statistically unusually high.

Especially surprising were the findings of a pilot study carried out in Alexandria to establish the frequency of congenital adrenal hyperplasia among newborn babies. Congenital adrenal hyperplasia (CAH) is by far the most frequent condition altering the sexual development of genetic females (46,XX). The bodies of people with CAH lack one of the enzymes necessary for the proper functioning of the adrenal glands, which as a result produce excessive androgen. Female babies with CAH have normal female reproductive organs but may present slightly to very male-looking external genitalia at birth, such as, for example, an enlarged clitoris that resembles a small penis. Without intervention, girls' bodies typically continue to virilize as they mature, and they may experience menstrual problems, a deep voice, and hirsutism.[4] The pilot study found the second-highest known prevalence of CAH in the world, 1 in 1,209 live births, after the Yup'ik Eskimos of Alaska (1 in 282). The worldwide rate is estimated at 1 case of CAH in 14,199 live births.[5] Many cases of CAH have been also reported in Jordan and in

38. Rispler-Chaim, on the other hand, calls the jurists' attention to hermaphroditism "astonishing and commendable" in her *Disability in Islamic Law*, 71.

4 To be precise, congenital adrenal hyperplasia is an umbrella term for a group of autosomal recessive disorders of the adrenal glands. CAH occurs in males (46,XY) as well. Male babies with CAH have male-looking genitals and reproductive organs, but the effects of the excess androgen can be felt as early as two or three years of age, when the boy appears to enter puberty. Certain forms of CAH, such as the "salt-losing" type (the total or almost total deficiency of the 21-hydroxylase enzyme results in salt and water loss), are dangerous and, if not treated promptly, can even lead to death. For more information about CAH and many other disorders of sex development, see the website of the Accord Alliance (www.accordalliance.org) and Catherine Harper's *Intersex* (Oxford; New York: Berg, 2007).

5 Shawky M. Tayel, et al., "Congenital Adrenal Hyperplasia in Alexandria, Egypt: A High Prevalence Justifying the Need for a Community-based Newborn Screening Program," *Journal of Tropical Pediatrics* 57, no. 3 (2011): 232–234. Other studies

Lebanon, and although there are no clinical studies of DSD in Syria, it is improbable that the results would differ much from those in the neighboring countries.[6]

The rare autosomal recessive inherited disorder 5-α reductase-2 deficiency (5-ARD) has also been documented throughout the Arab world. XY babies lack the enzyme 5-α reductase and are born with ambiguous or feminine-looking genitalia. They are thus usually raised as girls. Around the time of puberty their bodies masculinize, and as teenagers most make the sometimes painful transition to living as males. 5-ARD runs in families and is found in clusters only among certain populations, including in the Dominican Republic, in Papua New Guinea, and in the Middle East. One pediatric surgeon alone reported eighty-two cases of 5-ARD in Egypt, and a team of pediatric endocrinologists found eight members of a large extended family with 5-ARD in southern Lebanon, one of the largest such clusters discovered anywhere in the world. All eight members of this family were raised as females and changed their gender role to male after their bodies masculinized at puberty.[7]

on Egypt include I. Mazen, et al., "Screening of Genital Anomalies in Newborns and Infants in Two Egyptian Governorates," *Hormone Research in Paediatrics* 73, no. 6 (2010): 438–442; S. I. Ismail and I. A. Mazen, "A Study of Gender Outcome of Egyptian Patients with 46,XY Disorder of Sex Development," *Sexual Development* 4, nos. 4–5 (2010): 285–291; S. A. Temtamy et al., "A Genetic Epidemiological Study of Malformations at Birth in Egypt," *Eastern Mediterranean Health Journal* 4 (1998): 252–259.

6 For Jordan, see Hussein Al-Maghribi, "Congenital Adrenal Hyperplasia: Problems with Developmental Anomalies of the External Genitalia and Sex Assignment," *Saudi Journal of Kidney Diseases and Transplantation* 18, no. 3 (2007): 405–413; M. A. Arnaout, "Late-Onset Congenital Adrenal Hyperplasia in Women with Hirsutism," *European Journal of Clinical Investigation* 22, no. 10 (1992): 651–658. For Lebanon, see Valérie Delague et al., "Mutational Analysis in Lebanese Patients with Congenital Adrenal Hyperplasia Due to a Deficit in 21-Hydroxylase," *Hormone Research* 53, no. 2 (2000): 77–82.

7 Nabil M. Dessouky, "Gender Assignment for Children with Intersex Problems: An Egyptian Perspective," *Egyptian Journal of Surgery* 20, no. 2 (2001): 499–515. For references to more Egyptian clinical studies on 5-ARD and other DSD, see Samia A. Temtamy, Mona S. Aglan, and Nagwa A. Meguid, "Genetic Disorders in Egypt," in *Genetic Disorders*, ed. Ahmad S. Teebi, 245–247. For Lebanon, see Ze'ev Hochberg et al., "Clinical, Biochemical, and Genetic Findings in a Large Pedigree of Male and Female Patients with 5α-Reductase 2 Deficiency," *Journal of Clinical Endocrinology and Metabolism* 81, no. 8 (August 1996): 2821–2827. Incidentally, this condition became relatively well known following the publication of Jeffrey Eugenides's novel

Palestine is particularly interesting and perhaps even unique with respect to DSD. A very rare inherited autosomal recessive condition, 17ß-hydroxysteroid dehydrogenase deficiency-3, which affects the biosynthesis of testosterone in 46,XY fetuses (females can have 17ß-HSD3 deficiency, too, but they are asymptomatic), has been found among Palestinians living in Israel, the West Bank, and Gaza. The prevalence of 17ß-HSD3 deficiency in Jabalya, a town near Gaza, may be as high as 1 in 100 to 150, a staggeringly high frequency compared to the estimated worldwide average of 1 in 147,000.[8]

In the 1980s and 1990s researchers were able to trace the genealogical tree of an extended Muslim family with many individuals with 17ß-HSD3 deficiency over eight generations. They identified the family's earliest known ancestors, originally Druze who migrated to the Palestinian coast from the mountainous regions of Lebanon and Syria in the nineteenth century, and attributed the gene for the deficiency to them. Although the hormonal pattern is different in each type, genetic males with 17ß-HSD3 deficiency develop in a way similar to those affected by 5-ARD. All were born with ambiguous or female-looking genitalia, with a phallus resembling a normal or enlarged clitoris and with undescended, and thus undetected, testes. Consequently, the children were raised and socialized as females. On reaching puberty, as a result of complex hormonal processes, their bodies spontaneously virilized. The adolescents began to display beards and male bodily hair, their voices deepened, and their phalluses grew. After they masculinized, the majority switched to a male gender identity and a male gender role. Most of their family members more or less readily accepted their transition and they were apparently well integrated in society with their new male gender role. Some older individuals, on the other hand, had a male gender identity but chose to continue to live as females, even though both they and their families knew that they were genetic males. Others still were not given any choice in the matter, and, having

Middlesex, which won the 2003 Pulitzer Prize for fiction, and whose protagonist, Cal, has 5-ARD.

[8] In 2009 a Gaza-based pediatric urologist reported that he had personally come across some eighty cases in the course of seven years. Ivan Watson, "Rare Gender Identity Defect Hits Gaza Families," CNN, December 17, 2009. http://www.cnn.com/2009/WORLD/meast/12/17/gaza.gender.id/.Accessed January 28, 2012.

been castrated by their physicians shortly after their condition was diagnosed, continued to live as females.[9]

Interestingly, according to the authors of these clinical studies one of the Arabic words used by Palestinians to refer to people with intersex conditions is *mushkil*, which means "ambiguous" or "problematic." It is unlikely to be a coincidence that, as we are about to see, *khunthā mushkil* is exactly the word that has been used in Arabic since at least early Islamic times to describe the "ambiguous hermaphrodite."[10]

There is no doubt, then, that the high rate of consanguineous marriages plays a fundamental role in the frequency of DSD in the region. To cite one last example, a recent study found that the percentage of consanguinity in the families of 208 Egyptian patients with ambiguous genitalia was over 60 percent overall and as high as 72 percent for disorders that affect genetic females.[11]

Premodern *Khunthā*s

Precisely because of their inherited genetic component, there is no reason to believe that the frequency of many intersex conditions would have been substantially lower in Ottoman times. Of course, several conditions that today are labeled DSD would not have been identified

[9] Ariel Rösler, "Steroid 17ß-Hydroxysteroid Dehydrogenase Deficiency in Man: An Inherited Form of Male Pseudohermaphroditism," *Journal of Steroid Biochemistry and Molecular Biology* 43, no. 8 (1992): 989–1002; Ariel Rösler and Gertrude Kohn, "Male Pseudohermaphroditism Due to 17ß-Hydroxysteroid Dehydrogenase Deficiency: Studies on the Natural History of the Defect and Effect of Androgens on Gender Role," *Journal of Steroid Biochemistry* 19, no. 1 (1983): 663–674; G. Kohn et al., "Male Pseudohermaphroditism Due to 17ß-Hydroxysteroid Dehydrogenase Deficiency (17ßHSD) in a Large Arab Kinship: Studies on the Natural History of the Defect," *Journal of Pediatric Endocrinology* 1, no. 1 (1985): 29–37; Milton Diamond, "Pediatric Management of Ambiguous and Traumatized Genitalia," *The Journal of Urology* 162, no. 3 (September 1999): 1023.

[10] G. Kohn et al., "Male Pseudohermaphroditism," 35. The authors of the article translate "moshkel" (*mushkil*) as "troublemaker," which I believe is inaccurate. Another name used both in Gaza and in southern Lebanon for intersex individuals was "shakar" or "shakr," which the authors state is "a combination of the Arabic words for male and female." Ibid., and Hochberg, "Clinical, Biochemical, and Genetic Findings," 2821.

[11] I. Mazen, O. Hiort, R. Bassiouny, and M. El Gammal, "Differential Diagnosis of Disorders of Sex Development in Egypt," *Hormone Research* 70, no. 2 (2008): 118–123.

as such at a time when ultrasounds and blood tests did not exist. For instance, individuals with complete androgen insensitivity syndrome (CAIS), an almost always inherited DSD that affects 46,XY males, would have looked typically female and in most cases would have had a female gender identity. The fact that they did not menstruate or bear children would not have been attributed to their being genetic males (people with CAIS have breasts and a vagina, but no uterus or ovaries).[12]

The Arabic term for intersexed, *khunthā* (pl. *khināth* or *khanāthā*), is entirely unrelated to the Greek mythical figure of Hermaphrodite but derives from the Arabic root *kh-n-th*, which means "to be soft, to be effeminate." With minor variations, the same basic definition of khuntha appears in early modern legal compendia, fatwa collections, lexicons, biographical dictionaries, chronicles, and dream manuals: a person who has both a penis and a vagina (the Arabic word *farj* was used to mean both vulva and vagina, thus potentially creating some confusion when discussing female genital anatomy). Some added that khuntha could also refer to someone who possesses neither organ.[13]

[12] Androgen insensitivity syndrome (AIS) affects sex hormones. People who have this inherited syndrome do not respond at all, or respond only partially, to the androgen ("male") sex hormones such as testosterone, and thus their bodies appear more or less female depending on the degree of androgen insensitivity. If the androgen insensitivity is partial (PAIS), the external genitalia will look ambiguous and will present what looks like an enlarged clitoris or a very small penis. If the insensitivity is complete (CAIS) the overall appearance of the person conforms with female gendered norms, the external genitalia will appear female, with a vagina often shorter than average and blind ending, breasts will develop at puberty, and little body hair will be produced. AIS is often first suspected at puberty with the absence of menstruation, since people with this syndrome do not have a uterus or ovaries and therefore do not menstruate (they have testes, but they are internal and thus are usually undetected).

[13] Al-Zabidi, *Taj al-'Arus*, 5:240–243; al-Tumurtashi, *Matn Tanwir al-Absar*, 245; al-Halabi, *Multaqa al-Abhur*, 334; al-Nabulusi, *Ta'tir al-Anam*, 1:192; al-Muradi, *Silk al-Durar*, 3:278. The word *khunthā* is masculine or neuter, and thus the personal pronoun used is *huwa* ("he" or "it"), which can create confusion or ambiguity when translating the word into English. "Hermaphrodite" derives from Hermaphroditos, the mythical son of the Greek deities Hermes and Aphrodite. The most familiar version of the story of Hermaphroditos is recounted by the Roman poet Ovid (d. 17 CE) in Book IV of his *Metamorphoses*. Hermaphroditos, a youth of exceptional beauty, was one day sighted by the nymph Salmacis while he was bathing in her fountain. Overwhelmed by his beauty, Salmacis was instantly smitten with him and so ardently desired union with him that she pleaded with the gods to join them

Stories about khunthas were at times included under the heading of marvels (*'ajā'ib*) or oddities (*gharā'ib*), but they were generally not narrated with contempt or condemnation. The biographer Muhammad al-Muhibbi (d. 1699) relates several anecdotes about female-to-male sex change that he had gathered from various sources. For example, when the daughter of a Cairene prince turned fifteen in December of 1325, "her vagina closed up. A penis and testicles sprouted and she began to have nocturnal emissions like men do." The news spread throughout Cairo until it captured the attention of the powerful Mamluk ruler Amir Manjak, who summoned the girl so that he could ascertain the veracity of her story. Persuaded that she was in fact a man, he ordered her to abandon the women's clothes she was wearing in favor of the soldier's garb with which he presented her. After renaming her/him Muhammad, Amir Manjak "appointed him as one of his attendants and granted him a piece of land. And this was witnessed by everyone." A similar and perhaps even more spectacular case of sex change took place in Qisha, one of the provinces of Isfahan. A girl was given in marriage, and "on her wedding night, her pubic area began to itch and, on that same night, a penis and two testicles came out to her, and she became a man."[14]

The tone with which the stories of the two adolescent khunthas/girls who spontaneously turned male are told is infused with wonder, but not with revulsion or fear. In fact, I have located only one account of a khuntha that features an arguably negative association with its subject. 'Ata' Allah al-Mawsili (d. after 1727) was a pious and well-traveled Sufi shaykh from Mosul. During one of his journeys in faraway lands, possibly on the coast of Oman or India, al-Mawsili encountered a white-bearded shaykh. The shaykh, whose countenance he describes as "luminous," approached al-Mawsili. Thinking that he must be an important local figure, al-Mawsili rose to honor him and kissed his hand. Thereupon, one of the men present said to him: "Oh Shaykh, this man is a Magian!" After hearing these words, al-Mawsili sought God's

together. Taking her request literally, the gods joined the bodies of Hermaphroditos and Salmacis, creating one Hermaphrodite.

[14] Both stories are in al-Muhibbi, *Khulasat al-Athar*, 3:316. A third similar case is that of a girl by the name of Safiyya. When "she turned fifteen, a penis appeared to her and a beard grew out to her." Ibid.

forgiveness for having honored such a man. "Shall I tell you something even more amazing (*a'jab*) than that?" the speaker continued. "He is a khuntha, with two organs. First he was married to a man, and bore him a child; then he married a woman and fathered a child with her, too!"[15] In this instance, being a khuntha was "even more amazing" than being a Magian, that is, an adherent of Mazdaism (Zoroastrianism), an ancient Persian religion whose followers were commonly believed to be fire worshippers. There is no indication, however, that he was categorized as a monstrous being, or as a bad omen, or that he was viewed as a sign of God's wrath, as scholars have argued of hermaphrodites in early modern Europe.[16]

Medical Approaches to Ambiguous Genitalia

None of the Ottoman-era sources I have examined, medical or otherwise, offer any explanation for why some people are born khunthas in the first place, but in earlier times some Muslim physicians had attempted to account for the existence of khunthas with scientific theories of conception originating with Hippocrates (d. ca. 377 BCE), Aristotle (d. 322 BCE), and Galen (2nd c. CE). The central issue in these theories was the respective contribution of the male and the female to the embryo's creation. The sex of a fetus depended on several elements, including whose seed (the man's or the woman's) predominated and on which side of the uterus the seed fell. The womb, it was believed, was divided into seven cells or chambers, three on the left, three on the right, and one in the middle. If the embryo developed in one of the three right-hand cells, it would be male; if in one of the left-hand cells, female; and if in the middle cell, intersex.[17]

15 Al-Muradi, *Silk al-Durar*, 3:278.

16 See Ruth Gilbert, *Early Modern Hermaphrodites: Sex and Other Stories* (New York: Palgrave, 2002), and "'Strange Notions': Treatments of Early Modern Hermaphrodites," in *Madness, Disability and Social Exclusion: The Archaeology and Anthropology of "Difference,"* ed. Jane Hubert (London; New York: Routledge, 2000), 144–158.

17 See Sanders, "Gendering," 76; Basim Musallam, *Sex and Society in Islam: Birth Control Before the Nineteenth Century* (Cambridge: Cambridge University Press, 1983), chapter 3. See also Joan Cadden, *The Meaning of Sex Difference in the Middle Ages: Medicine, Science, and Culture* (Cambridge: Cambridge University Press, 1993), 198–199.

Surgeons offered practical advice on treatment. The Arab Andalusian physician Abu al-Qasim al-Zahrawi (d. 1013), better known in Europe as Albucasis, dedicated chapter 70 of his *On Surgery* to "the treatment of the hermaphrodite (khuntha):"

> There are two kinds of male hermaphrodite: one has the appearance as of female pudenda with hair in the region of the perineum; the other has the same in the skin of the scrotum between the testes. Sometimes also urine is excreted through that which is in the skin of the scrotum.
>
> There is also one kind among women, in which there is, above the female pudenda, on the pubes, what resembles the male organs. These are small indeed, but protuberant, one being like the penis and the two others like testicles.
>
> This is the treatment of all three kinds, that is, two male and one female: the superfluous growths must be cut away so that every trace is destroyed; then the usual treatment for wounds should be applied until they are healed. As to the second kind, in man, where the opening is in the scrotum and the urine runs out of it, there is no operation for this, and no cure.[18]

This passage shows that khunthas could be either male or female; that what made a person intersexed was the presence of both male and female *genital* tissue (there is no mention of other markers of femininity or masculinity such as breasts or beards); and that treatment consisted in the excision of the "superfluous growths."

On Surgery was one of the most celebrated and widely used surgical manuals throughout the Middle East as well as Europe in the early modern period. Albucasis's description of khunthas was also reproduced in an Ottoman Turkish surgical manual written by Şerefeddin Sabuncuoğlu (d. ca. 1468), a physician who practiced medicine in the hospital of Amasya, an important center of learning in north-central Anatolia. Sabuncuoğlu dedicated and presented his work, *Imperial Surgery*, to Sultan Mehmet II. The book is a translation from Arabic into Turkish of Albucasis's manual, with the occasional addition

[18] M. S. Spink and G. L. Lewis, editors and translators, *Albucasis on Surgery and Instruments: A Definitive Edition of the Arabic Text with English Translation and Commentary* (Berkeley and Los Angeles: University of California Press, 1973), 454–455. On the influence of Albucasis in the early modern Middle East, see Shefer-Mossensohn, *Ottoman Medicine*, 50–51. The Arabic title of his encyclopedic work is *Kitab al-Tasrif*. Volume 30 is dedicated to surgery.

of Sabuncuoğlu's own comments.[19] Most fascinating are the accompanying miniature paintings of many surgical procedures. The paintings illustrate not only the surgical instruments, as in the original *On Surgery*, but also female and male doctors operating on female and male patients of all ages.

What is evident from both manuals is that many genital and physical anomalies that today are likely to be considered intersex conditions were not labeled as such in the premodern period. Consider, for example, chapter 47, "On the treatment of the male breast when it resembles the female:" "The breasts of some males may swell on attaining puberty so as to resemble the female breast, and they remain permanently swollen and ugly. *If* this is abhorrent, a semicircular incision should be made on the breast [. . .]. Then dissect away all the fatty tissue and pack the wound with a cicatrizing compound and sew the lips together and dress until healed."[20] The corresponding entry in the Turkish manuscript includes a miniature painting of an older male doctor with a long white beard excising the female-looking breast of an adult man sporting a black beard (folio 89a). Although female-looking breasts on a man might have warranted cosmetic surgery, the sex of the man was simply not questioned. It is also noteworthy that physicians did not automatically assume that female-looking breasts on a man would necessarily be abhorrant or require treatment ("*if* this is abhorrent").

Chapter 55, "On the treatment of boys born with imperforate urinary meatus; or with the meatus small or not in the proper place," is most likely what modern medicine today calls hypospadias:

> Sometimes a boy is born from his mother's womb with the glans penis not perforate. So at the moment of his birth you should be quick and make a perforation with a fine scalpel [. . .]. In those cases having a misplaced meatus – for example, there are some born with the opening at the edge of the glans so

[19] There are three known manuscript copies of *Cerrahiyyetü'l Haniyye* (Imperial Surgery), two in Istanbul and one in Paris, France. I have used the Paris copy, Bibliothèque nationale de France, Département des manuscrits, Supplément turc 693 which can be read online in its entirety through the library's website at http://gallica.bnf.fr/ark:/12148/btv1b8427201w. Accessed January 30, 2012. Note that the catalog of the Bibliothèque nationale de France incorrectly lists this manuscript as being the translation of an original Persian work written under Mongol rule in Persia.

[20] *Albucasis on Surgery*, 362. Italics mine.

that they cannot make water forward until the opening of the penis is lifted up by hand, nor can they beget, since the semen cannot reach directly to the uterus, a disorder which is very ugly. The manner of operating in this case is for the patient to lie back; then with your left hand draw out his glans firmly and cut the end of the glans at the place of the meatus with a broad knife or sharp scalpel as if sharpening a pen, or as if you were whittling something, so that the middle of it protruded like a glans penis and the opening falls in the middle as it should.[21]

Chapter 72 describes how to treat an unperforated vulva (labial adhesions). Sabuncuoğlu's Turkish version includes an illustration of two female doctors, one of whom holds a younger woman still while the other doctor attempts to separate her labia manually (folio 112a).

But perhaps most striking of all is the first part of chapter 71 of Albucasis's *On Surgery*, "On cutting the clitoris and fleshy growths in the female genitalia." The translation from the Arabic of the first part reads: "The clitoris may grow in size above the order of nature so that it gets a horrible deformed appearance; in some women it becomes erect like the male organ and attains to coitus. You must grasp the growth with your hand or a hook and cut it off. Do not cut too deeply, especially at the root of the growth, lest haemorrhage occur. [. . .]" The Turkish version is sufficiently different to merit quoting it in full (folio 110b–111a):

There is a part of the female vulva that is called *tılak* (clitoris) in Turkish. In some women it is so big that it may be ugly to look at [*şöyle büyük kim nazarda kabih olur*] and in some women it is as big as the male member and they have intercourse like men. In Arab lands [*diyar-ı Arabda*] they cut it. The way to do it is to hold the redundant part that should be cut in your hand, or to hold it with an implement, and to pull it upwards, but do not cut off the skin so as not to create a blood flow [*ta'kim kan boşanmaya*].[22]

21 *Albucasis on Surgery*, 388–391. No hereditary link has been conclusively established for hypospadias, which may be caused by environmental factors. Today it is the most common DSD in males, with a prevalence as high as 1 in 150 births.

22 The translation from Arabic is in *Albucasis on Surgery*, 456–457; the translation from Turkish is by Dror Ze'evi in *Producing Desire: Changing Sexual Discourses in the Ottoman Middle East, 1500–1900* (Berkeley: University of California Press, 2006), 42. It is unclear if the miniature painting described above refers to chapter 70, about khunthas, or to chapter 71, on the removal of large clitorises. I tend to agree with Dror Ze'evi that it pertains to chapter 71.

This intriguing text, which implies that the practice of cutting off penis-like clitorises is an Arab practice, is accompanied by a painting of a female physician operating on an adult female patient, both of whom are wearing veils that cover their hair. Out of the vulva of the patient there protrudes what resembles a penis, depicted as very long, perhaps to convey that it is not merely an enlarged clitoris. The female doctor is portrayed grasping the penis in one hand while the other hand holds a surgical instrument with which she is about to excise the excess flesh.

The Law in Theory

In light of the above, it is not surprising that Islamic law dealt fairly extensively with intersex conditions. Jurists were not merely indulging in exercises in legal logic analogous to the proverbial Christian theologians' speculations about the number of angels that could dance on the head of a pin, but rather were responding to concerns grounded in reality. To be sure, it was not nearly as common a reality as blindness, deafness, or impairments of the mind, but it was a reality nonetheless.

Because of the societal responses they evoked, ambiguous genitalia could be highly disabling in the spheres of marriage and reproduction. Since many anomalies are not visible at birth and may not manifest themselves until puberty, oftentimes it was probably not until a boy or a girl was married off that any atypical genitalia would be detected. At that juncture they were a serious impairment if they prevented the consummation of marriage, which, defined explicitly and narrowly as vaginal penetration, was cardinal to the marriage contract. The physical inability to consummate a marriage on the part of either a husband or a wife could lead to its dissolution. In fact, although a man could divorce his wife at any time and for any reason, a husband's impotence, regardless of its cause, was one of the few grounds available to a woman for a marriage annulment (which then raised financial issues such as the repayment of the dower given to the bride by the groom). Being known as unable to consummate a marriage could also substantially curtail a person's future marriage and reproductive prospects. In societies in which almost everyone was expected to marry and to have children, this reputation could be severely damaging to an individual's social standing, especially if she was a woman.

Muftis and judges were well prepared to rule on these situations when called on to do so. In Islamic law male and female urogenital anomalies were labeled as defects or flaws (*'uyūb*) and were filed under the rubric of "physical impediments" to sexual intercourse. There were minor variations among the four main schools of law, but any condition that prevented vaginal penetration usually constituted a "defect." More specifically, female urogenital impairments included "when the only opening a woman has is the urethra" (*rataq* or *talāḥum*), probably labial adhesions, "a bone or something else that prevents the penis from penetrating" (*qarn*), and "a hernia on the vulva" (*'afal*). In men, they included having the penis completely cut off or both the penis and the testicles severed (*majbūb*). In both cases the wife would be granted the immediate annulment of the marriage, if she desired it, because there was no expectation that the husband would ever be able to have the penetrative sex to which she was legally entitled.[23]

The rules concerning impotence were slightly more complex. A separate chapter in law books and fatwa collections was often devoted to the impotent (*'innīn*). In the plain words of a seventeenth-century mufti from Aleppo: "In the law, [the impotent] is he who does not have sexual intercourse with a woman despite the presence of the member [penis], or only has sexual intercourse with the deflowered woman but not the virgin, or with some women but not with others. That is because of an illness, or because of a weakness in his constitution, or because of old age, or because of sorcery."[24]

The disabling category of "impotent" encompassed other conditions, including having a penis too short or too small to effect vaginal penetration, being a eunuch (*khaṣīy*), whose testicles but not the penis

[23] All the preceding definitions are taken from al-Kawakibi, *al-Fawa'id al-Samiyya*, 1:289–293. Note that the defects *qarn* and *'afal* may well have encompassed cases of undescended testes in people with 5-ARD or 17ß-HSD3 deficiency or may have been genital malformations due to any of several other DSD, such as partial androgen insensitivity syndrome or congenital adrenal hyperplasia.

[24] Al-Kawakibi, *al-Fawa'id al-Samiyya*, 1:289–293. See also al-Tumurtashi, *Matn Tanwir al-Absar*, 80; al-Haskafi, *al-Durr al-Mukhtar*, 4:249–250, 5:165–176; Ibn 'Abidin, *Radd al-Muhtar*, 4:249–250, 5:165–176. For fatwas concerning impotence, see al-Husayni, *Fatawa*, 17b–18a, and 20a; al-Ramli, *al-Fatawa al-Khayriyya*, 1:56; al-'Imadi (as edited by Ibn 'Abidin), *al-'Uqud al-Durriyya*, 1:33; Ibn al-Ha'ik, *Shifa' al-'Alil*, 14; al-Hanuti, *Muntakhab al-Fatawa*, 36a-b; al-Tarabulusi, *Fatawa*, 63. See also Rispler-Chaim, *Disability in Islamic Law*, 53–57.

had been removed, and being a *shakkāz*, defined as someone who ejaculates before penetrating a woman. The impotent man was granted one year to attempt to have intercourse, after which, if still unsuccessful, the wife could obtain a divorce.

But most disabling of all was being labeled khuntha. Blind, deaf, and deaf-mute people were only legally disabled in certain circumstances (testimony in court proceedings, for instance), but khunthas were disabled in most contexts. Law books and sometimes fatwa collections contained a section titled *Kitāb al-Khunthā* ("The Book of the Khuntha"). Jurists offered no explanation, biological, religious, or otherwise, for the existence of khunthas. They apparently accepted the possibility of sex indeterminacy as an undeniable, if perhaps inexplicable, fact of life. A khuntha was a legal persona, and the jurists' approach to the manifold issues surrounding intersex was, as usual, pragmatic. Consider the following passage by Ibrahim al-Halabi (d. 1549/50), one of the most prominent Hanafi legal scholars in Ottoman times:

It is one who has a penis and a vagina. If he urinates from one of them he is regarded to be of that sex. If he urinates from both organs, the organ from which the urine comes out first determines his sex. If he urinates from them simultaneously, then he qualifies as ambiguous (*mushkil*). The quantity of urine is not relevant. [. . .]

When he reaches puberty, if he displays some of the signs of manhood, such as the growth of a beard, the ability to have sexual intercourse or nocturnal emissions like a man, then he is a man. If some of the signs of womanhood appear, such as menstruation, pregnancy, breasts, the ability to breastfeed, or the ability to be penetrated during sexual intercourse, then he is a woman. If none of these signs appear, or if they are contradictory, then he is ambiguous (*mushkil*). Muhammad [al-Shaybani, (d. 804), one of the disciples of Abu Hanifa and the eponymous founder of the Hanafi school of law.] said: Ambiguity occurs before puberty, for when he reaches puberty there is no ambiguity.

When it is determined that a khuntha is ambiguous, the most cautious course is to be taken. He will pray veiled and, during prayer, will take his position between the rows of men and women. If he stands in the row of men, then the men who are immediately next, in front, and behind him must repeat their prayers [because their prayers would be invalid]. If he stands in the row of women, he has to say his prayers again.

He must not wear silk or jewelry. While in the state of ritual consecration during his pilgrimage to Mecca (*iḥrām*) he must wear a sewn garment [like women do]. He must not uncover himself in front of either men or women. He should only be alone with unmarriageable male and female relatives, and must not travel without an unmarriageable relative (*maḥram*).

Neither man nor woman can circumcise him. A female slave must be bought at his expense to circumcise him. If he cannot afford it, the state will pay for the slave and then sell her again.

If he dies before his sex is clarified, he is not to be washed [because of the possibility that the person washing him may be looking at the genitals of someone of the opposite sex] but rather cleansed with soil or sand (*tayammum*), and his body is to be covered with five winding sheets [as is the custom for women]. When he is a prepubescent adolescent (*murāhiq*), he should not attend the washing or the cleansing with sand of a deceased man or woman.... During a collective funerary prayer, he will be placed between men and women.

As for inheritance, he will receive the smaller of the two possible shares of an estate. If his father dies leaving him and a brother, the brother will inherit two shares and the khuntha one. According to al-Sha'bi [a trustworthy Hadith transmitter upon whom Hanafi jurists often relied] he is owed half of the two shares....

If he is a slave and his master declares "all my male slaves are free" or "all my female slaves are free," the khuntha mushkil will not be manumitted until his state is clarified.

If, after it has been ascertained that he is an ambiguous hermaphrodite, he says "I am a man" or "I am a woman," his statement will not be accepted. Before that, however, his statement will be accepted.[25]

This excerpt indicates in no uncertain terms that the jurists' principal concern was to attempt to place the khuntha, defined according to purely genital criteria as an individual who possesses both a penis and a vagina, within the category of either man or woman. This preoccupation was driven less by moral concerns than by a host of practical legal issues. To what share of an inheritance was a khuntha, and, consequently, the other members of his/her family, entitled? Whom could a khuntha marry? How could s/he perform her/his basic religious duties, such as participating in the Friday communal prayer? Could a khuntha

[25] The passage is taken from al-Halabi, *Multaqa al-Abhur*, 334–336. For similar discussions of khunthas in other Ottoman-era law compendia, see al-Tumurtashi, *Matn Tanwir al-Absar*, 245; al-Haskafi, *al-Durr al-Mukhtar*, 10:446–452; al-Haskafi, *al-Durr al-Muntaqa*, 4:467–472; Ibn 'Abidin, *Radd al-Muhtar*, 10:446–452; al-Kawakibi, *al-Fawa'id al-Samiyya*, 2:484–486; Ibn Nujaym, *al-Ashbah wa-l-Naza'ir*, 3:371–380; al-Hamawi, *Ghamz 'Uyun al-Basa'ir*, 3:371–380.

testify in court? Who should circumcise or wash the body of a deceased khuntha? These were all important matters that affected not only the individual in question but also the family and, more broadly, the community in which s/he lived. It was therefore imperative for jurists to make a vigorous attempt to assign a definite sex, or at least a definite gender, to the khuntha.

What guidelines should govern this attempt? As is clear from al-Halabi's passage above, if ambiguous genitalia were observed *before puberty*, the prescribed procedure for determining the child's sex entailed identifying the primary urinary orifice. The rationale for this criterion, which dates back to pre–Islamic Arabia, is that the primary function of the genitals when a child is born is to urinate, and that the sexual and reproductive functions become relevant only at puberty. Following its endorsement by the Prophet Muhammad as recorded in the Hadith literature, this standard for determining the sex of a child with ambiguous genitalia was incorporated into Islamic law. If a prepubescent khuntha had two urinary orifices and urinated from both simultaneously, then the sex of the child would remain in question, and s/he would be classified as *mushkil* or "ambiguous" (the term that, as mentioned earlier, is employed in Gaza to this day to refer to people with intersex conditions).[26]

If jurists did not elaborate further on the issue of determining the sex of a child, it was because, except in matters of inheritance, it was not of the utmost importance until a person reached the socially critical stage of puberty. The need to make a vigorous attempt to place a khuntha within the category of either man or woman acquired more urgency precisely when an individual attained puberty, because it was then that an individual assumed all the rights and obligations that accompanied adulthood, including the right to marry and divorce, the right to own and manage property, and the right and duty to participate fully in religious ritual. Once that stage was reached, the law instructed jurists to seek specific indicators to determine the khuntha's sex: the appearance of a beard, breasts, menstruation, pregnancy, nocturnal emissions, and the role of penetrator or penetrated in sexual intercourse.

[26] See Ibn 'Abidin, *Radd al-Muhtar*, 10:446–447. See also Cilardo, "Historical Development," 128–132, and Sanders, "Gendering," 77.

There were cases, however, in which no markers of masculinity or femininity surfaced, or they were contradictory: the simultaneous appearance of beard and breasts, or ejaculating from the organ of men and menstruating from the organ of women, or urinating from the organ of women and ejaculating from the organ of men.[27] In those problematic scenarios, the adult khuntha was labeled "ambiguous" (*mushkil*) and assumed an awkward legal status in between that of a man and a woman. In matters of inheritance, for example, some Hanafi jurists ruled that a khuntha mushkil should receive the share of a female, on the basis that the lesser share is the least to which either sex is entitled to by law. Other Hanafi jurists, however, argued in favor of a special share that consisted of half the share normally due to men plus half the share normally due to women, thus highlighting the "in-between" status of intersex the person.[28]

As for religious ritual, when praying in a mosque a khuntha mushkil was required to stand in a space between the rows allotted to men and the rows allotted to women. The logic was that, were the khuntha later to be declared a woman, she would simply have stood in the first row of women, but were he really a man, he would simply have stood in the last row of men. Thus, gender hierarchies would have remained unaffected. Significantly, not only were khunthas allowed to participate in communal prayer, but they could also officiate as imams by leading women in prayer (but not men, because men can only be led in prayer by men).[29]

Who was empowered to decide if an adult khuntha was really a man, really a woman, or "ambiguous"? The range of opinions among jurists ran the spectrum. A minority opinion argued that a khuntha's statement concerning his/her sex cannot be accepted under any circumstances. Implying that the highly disabling designation of khuntha mushkil should be left as the last resort, the majority contented that

[27] Ibn 'Abidin, *Radd al-Muhtar*, 10:447.

[28] The Hanbali and Maliki schools subscribed to this second opinion, on the grounds that "since the hermaphrodite is in an ambiguous position between a male and a female owing to his uncertain sex, one must apply the norm relative to males as well as that relative to females" (Cilardo, "Historical Development," 136). The laws of inheritance are meticulously detailed in Ulrich Rebstock's "Mathematische Quellen zur Rechtsgeschichte."

[29] Ibn 'Abidin, *Radd al-Muhtar*, 2:321.

the khuntha's declaration of his or her own sex was legally valid, on the grounds that no one would know one's sex better than the person in question.[30]

The Law in Practice

At the prescriptive level, Islamic law made serious efforts to mitigate the potentially disabling consequences of intersex for the person, the family, and the community, notably by permitting a khuntha to decide his or her sex. One wonders, however, to what extent the theory of the law was applied to real-life situations. Before delving into this crucial question, four notes of caution are in order.

First, the Ottoman-era narratives of intersex that are available to us are likely to be the exceptional ones, the complicated or contentious ones, the ones that came to the attention of judges, muftis, rulers, or to other members of the educated elite who, for whatever reason, chose to write them down.

Second, we do not have any first-person accounts like the memoirs of Herculine Barbin, the nineteenth-century French hermaphrodite made famous by Michel Foucault, and we will never know how many cases of intersex were not recorded, are lost, were undetected by families and communities, or were quietly resolved through the khuntha's declaration of her or his own sex or through family negotiations.

Third, in Egypt, the only country in the Ottoman Arab world where female genital cutting (FGC, also known as "female circumcision" and "female genital mutilation") has historically been practiced widely, male children mistaken for females at birth and consequently raised as girls might have had their penises inadvertently excised through FGC.

Fourth, it is impossible to gauge how many men and women continued to live with the gender role of the sex that they were assigned at birth. Contemporary clinical studies of intersex conditions in the Arab world show some resistance to sex reassignment from male to female and more readiness to switch from female to male. A recent Jordanian study revealed that the parents of seven out of fourteen genetic female children with congenital adrenal hyperplasia who had been

[30] Ibn 'Abidin, *Radd al-Muhtar*, 10:449–450. See also Cilardo, "Historical Development," 134–135.

misdiagnosed as males at birth because of severely ambiguous genitals decided that the children should remain male, even though it meant that they could never have children and might have a discordant gender identity. The uteruses and ovaries of six out of seven children were surgically removed.[31] The parents' decision can be partly explained by a preference (albeit by no means universal) for boys in societies in which males have greater economic opportunities and political and legal rights. Although it is always perilous to project current societal norms and behaviors onto the past, given that the scholarship on gender in the Ottoman Empire has conclusively demonstrated that women were indeed at a legal, political, and economic disadvantage, it is reasonable to presume that at least some khunthas would have chosen, when possible, to maintain a male gender role even if they felt that they were females, or that they would have been pressured by their families to do so.

These important qualifications notwithstanding, the compelling narratives of khunthas that have survived shed much light both on how the law was applied and on how intersex was culturally constructed. For example, Shaykh Salih al-Tumurtashi (d. 1645/6), a Hanafi mufti in Gaza, was asked to issue a ruling concerning the following dispute. Two men, identified by the standard pseudonyms "Bakr" and "Zayd," each fathered a child identified as a khuntha. While the children were still minors, Zayd married off his khuntha to Bakr's khuntha, but when they grew up the husband turned out to be a woman and the wife a man. We are not told, unfortunately, how those conclusions were reached, or what the markers of masculinity and femininity were in this particular case, or who made that determination.

The problem facing the mufti was to determine the legal validity of the marriage. Did the fact that the husband had become the wife balance out the wife's becoming the husband? If that were the case, would the marriage still be valid? After carefully perusing several legal sources, al-Tumurtashi apparently came to the conclusion that the laws pertaining to the marriage of khunthas were not only rather

31 Al-Maghribi, "Congenital Adrenal Hyperplasia," 408. Follow-up sessions indicated that only two out of seven appeared to have no difficulties with their forced male gender identity. For Egypt, see Dessouky, "Gender Assignment."

complicated but also somewhat contradictory. He decided to write a letter to the most prominent and well-respected mufti of seventeenth-century Palestine: Khayr al-Din al-Ramli. In the letter, al-Tumurtashi explained the issue at hand and asked his colleague, whom he addressed respectfully as the "solver of problems," to carry out "the utmost investigation of the matter." Al-Ramli obliged. After an exposition of the different opinions contained in the main works of Hanafi jurisprudence and fatwa collections, al-Ramli delivered his final verdict: the marriage was valid. Although the two spouses were khunthas, he explained, the legality of the marriage was technically suspended until the minor husband and wife reached puberty, and, with it, the removal of the ambiguity of their sex (and, the mufti adds, the two fathers should be reprimanded for marrying off their minor khunthas in the first place. They ought to have waited until they reached puberty, when their sex would become clear). But once it was confirmed that the two spouses were in fact of different sexes, there were no legal grounds to invalidate the marriage.[32]

Another time, al-Ramli was posed a question about inheritance, a practical matter that, like marriage, affected not only the khuntha but also other family members.

Question: About a khuntha who died, and about the fact that she was alleged to be female by a person who has a right to an allotted share of the khuntha's inheritance on the basis of the khuntha's being female. His evidence is that the khuntha used to urinate from the urinary orifice of women. Should his case be heard and his evidence accepted? And if you say yes, how should it be heard and accepted? And what about the *Hidaya*'s [by the 12th-century Hanafi jurist 'Ali al-Marghinani] prescription that if a khuntha dies before his status becomes clear he is not washed but rather his body is to be cleansed with soil or sand (*tayammum*) for precaution, and neither men nor women should look at him? And how can the khuntha's sex be proven, especially if the witnesses' statement that "we saw that she urinates from the urinary orifice of women" cannot be heard inasmuch as they do not meet the legal requirements of righteousness?

[32] Al-Ramli's *al-Fatawa al-Khayriyya*, 2:207–208. For a similar discussion of the rules regulating the marriage between two khunthas, see Ibn 'Abidin, *Radd al-Muhtar*, 4:61. Salih al-Tumurtashi, the mufti of Gaza to whom the request for a fatwa was originally posed, was the son of the illustrious Muhammad al-Tumurtashi (d. 1596), the author of the famous work of Hanafi jurisprudence *Tanwir al-Absar*.

This was a complex case made even more troublesome by the fact that one of the parties, perhaps a brother, clearly had a vested interest in assigning a female sex to his khuntha relative. Al-Ramli again set out to peruse several books of law. In a detailed fatwa he compared the sometimes contradictory opinions of various Hanafi jurists. In the end, he ruled that the statement of the witness who claimed that he saw the khuntha urinating like a female was valid (especially, we infer, since there were no witnesses stating the opposite), and that therefore the deceased khuntha should be treated as a female for the purposes of inheritance and burial.[33]

Al-Ramli followed the rule of the urinary orifice because the khuntha had died before reaching puberty. But what criteria ought to be followed when assigning a sex to an ambiguously sexed adult? A fatwa about a "man with two organs" provides some answers.

Question: There is a man (*rajul*) who has two [genital] organs (*alāt*): the organ of men and the organ of women. But the organ of men is blocked and nothing comes out of it when he urinates or has nocturnal emissions. Instead, his urine and semen come out of the hole (*thuqb*). His beard has appeared. Is he male, to be given the treatment of males in legal matters, or female, to be treated as such?

Here again we have a situation fraught with problems. As noted above, the law dictated that in determining the ambiguous sex of a post-pubescent individual the physical markers of masculinity and femininity should be sought. The obvious option with regard to a man possessing contradictory features, such as a beard, an ill-functioning sexual organ, and a female-seeming urinary orifice was to designate the man as a khuntha mushkil. Al-Ramli, however, exercised a considerable degree of latitude in interpreting the law and elected to render this ruling instead:

Answer: The situation being this, he is male, to be treated like males. The author of *al-Tatarkhaniyya* [a 14-th-century compendium of Hanafi law] has stated that ambiguity only occurs before puberty. As for after puberty, ambiguity is removed, because after puberty it is inevitable that there be a sign by which it is known that he is a man or a woman. If he has sexual intercourse with his penis he is a man. Likewise, if he does not have sexual intercourse

[33] Al-Ramli, *al-Fatawa al-Khayriyya*, 2:205–207.

with his penis but his beard has come out he is a man. And if he has nocturnal emissions like men do he is a man. And it should not be said that the exiting of his semen from the hole (*thuqb*) and the coming out of the beard are contradictory signs because of the probability that the canal of the penis is blocked. There is no contradiction. And God knows best.[34]

Al-Ramli's answer appears to be informed by the following rationale: why complicate matters unnecessarily, when the person in question has been living like an adult man and is clearly recognized as such (*rajul*)? There may be a simpler explanation for the ambiguity of the genitals of an otherwise male-looking individual with a beard, such as the "canal of the penis" being obstructed. The mufti chose to brush aside the evidence of contradictory signs of maleness and femaleness and chose instead to focus on the principle that "ambiguity only occurs before puberty." In other words, because Hanafi legal theory on sex ambiguity was itself somewhat ambiguous, it was within al-Ramli's purview to opt for the least problematic verdict and avoid creating a potentially unpleasant situation for an adult who up to that point had been living as a male.[35]

The Stories of 'Ali/'Aliyya and of Muhammad b. Salama and His Wife

The latitude enjoyed by Ottoman muftis and judges in interpreting the law in order to accommodate individual circumstances and their (culturally constructed) "common sense" approach is displayed nicely in the resolution of the story of 'Ali/'Aliyya in sixteenth-century Damascus, with which this chapter opened. To be sure, this fascinating story raises several questions to which we do not have the answers. We wonder, for example, why the judge decided to summon the young bookbinder to court in the first place. The text hints that something questionable had happened between the boy and the older suitor, but

34 Al-Ramli, *al-Fatawa al-Khayriyya*, 2:208.

35 Jurists did not always take al-Ramli's approach. In 1546 or 1547, a Damascus-based Maliki jurist reports making a stop in the southern Turkish town of Konya on his way to Istanbul. In Konya he saw a grown man "with a big beard flowing down to his chest," who was involved in trade. The reader is not told any of the details, except that it turned out that he was a woman and had "a female vulva." The town's ruler had him examined and found him to be a female, so he ordered him to shave his beard and start wearing the veil. She later married and got pregnant. In al-Muhibbi, *Khulasat al-Athar*, 3:315–316.

we are not told what. Perhaps their relationship was beginning to attract too much attention, or perhaps ʿAbd al-Rahman requested that the judge look into the matter, in the hope that ʿAli would be proven to be a woman, so that he could marry her, or perhaps ʿAli's family or he himself asked that he be examined. We also do not know if ʿAli consented to the excision of the protuberance covering his vulva.

Despite these lacunae, the story offers us another illustration of how the issue of switching sex and gender was dealt with in practice. ʿAli was a beardless youth (*shabb amrad*). In the context of the life cycle, the term "beardless youth" was used to refer to an adolescent boy in his teens who had not yet achieved complete physical maturity, symbolized by the most visible marker of maleness, the beard. In other words, ʿAli was at an intermediary stage between childhood and adulthood. If, in pondering his decision, the judge had followed the rules spelled out in Islamic legal doctrine, he should technically have sought the basic determinant of sex in prepubescent khunthas, the urinary orifice. If ʿAli was deemed to have reached puberty and there was no beard, the markers to consider should have been breasts, menstruation, or nocturnal emissions. There is no indication, however, that any of these criteria were a factor in the judge's assignment of female sex to ʿAli. In fact, in ʿAli's case the judge would have been equally justified in rendering the opposite ruling, designating ʿAli to be the male that he was brought up to be. A third option would have been to affix to ʿAli the disabling label of "khuntha mushkil." Instead, after calling on physicians to examine the youth's body and report on their findings, the judge opted for the solution that facilitated the ostensibly happy resolution of a situation laden with the potential for social disruption: the removal of the extra bit of flesh, the "small nipple" (which was possibly a small penis or a large clitoris). This was followed by the declaration that ʿAli was to be renamed ʿAliyya, and that she could now therefore marry her suitor.

Unlike the story of ʿAli/ʿAliyya, the account that follows did not end well. Najm al-Din al-Ghazzi (d. 1651) dedicated an entry of his biographical dictionary to Muhammad b. Salama, described as a pious Sufi from Damascus. At the beginning of the year 1506, while living in Cairo, this Muhammad married a woman described as being "clearly a khuntha" (*imra'a khunthā wāḍiḥ*). He consummated the marriage and "removed her virginity." Complications soon ensued, however, for Muhammad's wife had a paternal first cousin from the Maghrib who wanted to marry her, and thus had an informal right of first refusal. She turned down his request, so the cousin went to Amir Tarabay, one of the Mamluk chiefs in Egypt, to lodge a complaint against the couple. As a result of the cousin's complaint, the Amir had Muhammad and his wife brought before him. We are not told what transpired between the couple, the cousin, and the Amir, but we do know the outcome: the Amir ordered that the couple be severely flogged and be paraded through town in disgrace. Muhammad died from the injuries he sustained during this ordeal as soon as he reached the gate of the prison. "There is no power but in God," a sympathetic al-Ghazzi concludes, "and people felt very sorry for him."[36]

The renowned chronicler Ibn Tulun (d. 1546) relates the story of the hapless Muhammad with very different details and infuses it with a strikingly different moral. In early 1506, Ibn Tulun recounts news of "an amazing affair" (*amr 'ajīb*) that had recently occurred in Cairo. It involved Muhammad b. Salama al-Nabulusi, a young man who several years prior to the event had made a name for himself in Damascus as a member of a Sufi order. Muhammad then left Damascus for Cairo, where he joined other members of the order and began to congregate with beardless youths, as had been his custom in Damascus. Ibn Tulun, who is clearly suspicious of such congresses, hints at an unfavorable turn of fortune for Muhammad by warning the reader that "God wished to show his [Muhammad's] true nature." One day, Muhammad appeared in a Cairo court with a person wearing a veil and women's clothes, and asked to marry her. His request was approved. A few days later, some of his neighbors started to stir up trouble by suggesting, we infer, that Muhammad had actually married a boy. Those who had acted as witnesses to the marriage became fearful, probably because, had their testimony been proven to be false, they would have been subjected to judicial punishment. They informed Amir Tarabay of the situation. The Amir called in Muhammad's wife and launched an

[36] Al-Ghazzi, *al-Kawakib al-Sa'ira*, 1:51. Al-Ghazzi draws this account from *Hawadith al-Zaman wa-Wafayat al-Shuyukh wa-l-Aqran*, a chronicle by the Damascene scholar Ahmad Ibn al-Himsi (d. 1527 or 1528), who was living in Cairo at the time of the event. The historian 'Abd al-Hayy Ibn al-'Imad (d. 1679) relates the exact same version of the story (also drawn from Ibn al-Himsi) in *Shadharat al-Dhahab fi Akhbar Man Dhahab* (Beirut: al-Maktab al-Tijari li-l-Tiba'a wa-l-Nashr wa-l-Tawzi', 1966), 8:55.

investigation, at the conclusion of which the wife was found to be a boy dressed as a girl. She, however, professed to be a khuntha. To verify her claim, expert women witnesses were summoned to conduct a physical examination since she claimed to be female (note that in the story of ʿAli/ʿAliyya, male physicians were asked to inspect his body because ʿAli was, at that time, still considered a boy). The women concluded that he was undoubtedly male. At that point, according to Ibn Tulun, perhaps in a desperate attempt to prove her/his case "he wounded the base of his penis and claimed that he was menstruating." The women then inspected him more closely and again concluded that he was lying, which led Amir Tarabay to order that he be subjected to severe corporal punishment, paraded through Cairo on a bull, beaten again, and then sent to prison, where he died. For Ibn Tulun, the moral of the story is that "people started to believe even less" in Sufis like Muhammad.[37]

The two renditions of the tragic story of Muhammad and his wife differ dramatically. Ibn Tulun makes no mention of the cousin from the Maghrib and focuses his narrative on the deception perpetrated by Muhammad, who pretended to be a pious Sufi but was actually a morally corrupt individual engaged in an illicit marriage with a boy. This is an unlikely interpretation, for there would have been no need for Muhammad to take the extreme and dangerous action of marrying the boy if he simply wanted to maintain a relationship with him.

These divergent readings of the story may be explained by the authors' respective attitudes toward specific Sufi orders. Ibn Tulun was weary of Muhammad's Sufi order, which he accused of encouraging reprehensible practices such as congregating with beardless boys. Al-Ghazzi, on the other hand, takes a more sympathetic approach and it is clear that he does not believe that Muhammad acted inappropriately. Al-Ghazzi does not express any skepticism about Muhammad's piety, nor does he give any credence to claims that Muhammad's wife was a beardless boy disguised as a girl. Muhammad was a pious Sufi, and his wife was clearly a *female* khuntha, a fact corroborated by the statement that "he removed her virginity." Muhammad and his wife were killed as a result of malicious meddling by his wife's cousin,

[37] Ibn Tulun, *Mufakahat al-Khillan*, 1:297–298.

who felt entitled to marry her. For al-Ghazzi, then, there was a miscarriage of justice.

As interesting as the two historians' antithetical accounts of this tale are, and its veracity aside, the most salient point for our purposes is that each version not only accepts the possibility that Muhammad's wife could have been a female khuntha, but also acknowledges that, had that been the case, no illicit act would have taken place. Al-Ghazzi's disclosure that the female khuntha's paternal cousin was determined to wed her testifies to her desirability as a marriage partner, despite her status as "clearly a khuntha" (*khunthā wāḍiḥ*). Even Ibn Tulun, who obviously believes her to be a male impostor, implies that had she in fact been a khuntha there would have been no legal grounds to punish either spouse. The Mamluk chief apparently shared this opinion, for he gave her the benefit of the doubt and the opportunity to substantiate her claim by means of a physical examination. Perhaps most importantly, the girl herself chose to claim to be a khuntha, knowing that it might save her life.

The Roots of Sex Difference: Anatomy and Physiology

To recapitulate, khunthas were frankly acknowledged as a physical reality, and if a case of ambiguous sex reached the judicial authorities, serious attempts were made to accommodate that person's presence in society. These attempts invariably entailed assigning a definite sex to the khuntha and were informed by practical considerations, especially in matters that affected entire families, like marriage and inheritance. Individuals with ambiguous genitalia could traverse the sex divide through a process that was sanctioned by jurists and could entail, but did not necessarily require, a physical examination by doctors or midwives. Notably, little or no religious or moral stigma was attached to khunthas. They were not usually portrayed as monstrous beings or as the product of sin or divine displeasure. There were no public calls for their punishment or expulsion.[38]

38 A similar observation is also made by Rispler-Chaim, *Disability in Islamic Law*, 74. Scholarship on hermaphrodites in early modern Europe and in early America, on the other hand, has observed that intersex individuals were often categorized as monsters or regarded as a sign of God's wrath. For post-Reformation England, see Gilbert, "Strange Notions," and Reis, *Bodies in Doubt*, 3, for early America. On the concept

Among the factors that help explain why ambiguously sexed bodies were thus perceived is the permeability of sex boundaries in the Ottoman Arab world. Ideas about sex difference were very different from those expounded in the eighteenth and nineteenth centuries by European scientists, who endeavored to document sex difference in every part of the human body, especially through skeletal and cranial measurements. They are even further removed from today's hormone-based theories of biological differences. In the early modern Arab-Islamic world women were not fundamentally different from men in anatomy and physiology. The male and female bodies were not different in kind but in degree. In his groundbreaking *Making Sex: Body and Gender from the Greeks to Freud*, Thomas Laqueur argues that in Europe it was only at the end of the eighteenth century or early nineteenth century that bodily sex was definitively bifurcated, moving from a one-sex model to a two-sex model. It is only starting with the Enlightenment, according to Laqueur, that the current view that "there are two stable, incommensurable, opposite sexes and that the political, economic, and cultural lives of men and women, their gender roles, are somehow based on these 'facts'" became predominant. Before then, a monological rather than binary reading of sex entailed that "differentiations of gender preceded differentiations of sex," and that "to be a man or a woman was to hold a social rank, a place in society, to assume a cultural role, not to *be* organically one or the other of two incommensurable sexes."[39]

Although the persuasiveness of Laqueur's argument for European history is still being debated, Dror Ze'evi contends that this model, which he renamed "the imperfect-man model," was prevalent in the premodern Middle East. Anatomy was based largely on the writings of the ancient Greek physician Galen (2nd c. CE), who grounded

of the hermaphrodite as monster, see also Julia Epstein, "Either/Or – Neither/Both: Sexual Ambiguity and the Ideology of Gender," *Genders* 7 (Spring 1990): 107–108.

39 Thomas Laqueur, *Making Sex: Body and Gender from the Greeks to Freud* (Cambridge, MA: Harvard University Press, 1990), 6, 62, and 8, respectively. Critics of Laqueur's thesis have accused him of overstating his case and argue instead that the one-sex model was by no means hegemonic before the eighteenth century. The most vitriolic review of *Making Sex* is Katharine Park and Robert A. Nye, "Destiny Is Anatomy," *The New Republic* 204 (February 18, 1991): 53–57. Laqueur refuted some of his critics (persuasively, in my opinion) in "Sex in the Flesh," *Isis* 94 (2003): 300–306.

his arguments in observations drawn primarily from the dissection of animals. The illustrated anatomical work of Andreas Vesalius (d. 1564), *De Humani Corporis Fabrica*, which relied on human dissection, was gradually incorporated into Persian and Turkish medical texts in the form of anatomical drawings during the seventeenth century, and later into Arabic. Vesalius's contribution, however, did not substantially alter the Galenic conception of sex difference.[40]

Galen's anatomical system was incorporated into Arab-Islamic medicine through foundational authors like Avicenna (d. 1037). It acknowledged no fundamental structural differences, skeletal or otherwise, between the male and female anatomy. In fact, there was no "female body" as such. According to this "one-sex" or "imperfect-man" model, the female and male reproductive organs were regarded as counterparts of each other: the vagina was an interior penis, the womb an interior scrotum, the labia were the foreskin, and the ovaries interior testicles. There was no Arabic word reserved exclusively for "ovaries." Two common terms for testes, *al-bayḍatān* ("the two eggs") and *al-unthayān* (derived from the same Arabic root that means "to be female or feminine"), each applied to both male and female testes, and both were believed to produce semen necessary for conception.

The writings of the Egypt-based physician Da'ud al-Antaki (d. 1599) are instructive. Conforming to the Galenic paradigm he contended that the male's testicles were located externally, whereas the female's were smaller and located internally on each side of the womb. It is telling that the word *farj*, although most commonly used to refer to a woman's vulva or vagina, was also occasionally employed to indicate the male genitals.[41]

The similarities between the male and female bodies are visually represented in rare drawings of the human body contained in a treatise on anatomy by Shams al-Din al-'Itaqi (fl. 1632). Al-'Itaqi, a physician

40 Ze'evi, *Producing Desire*, especially chapter 1, "The Body Sexual: Medicine and Physiognomy," 16–47. For an overview of anatomy in Arab-Islamic medicine, see Emilie Savage-Smith, "Tashrih," *Encyclopaedia of Islam*, 2nd ed., 10 (2000): 354–356.

41 Al-Antaki, *Tadhkirat Uli al-Albab*, 3:36; al-Antaki, *al-Nuzha al-Mubhija*, 1:140–143; al-Zabidi, *Taj al-'Arus*, 5:242; Ibn 'Abidin, *Radd al-Muhtar*, 10:446. In the story of 'Ali/'Aliyya recounted above, for example, 'Ali was found to have a *farj unthā*, a "female vulva."

who resided in Istanbul during the reign of Murad IV, depicted the details of the female urogenital system in a way that removes any doubt that the uterus and vagina were conceptualized as the inversion of penis and scrotum.[42]

As noted in previous chapters, in the early modern period human physiology was understood within the framework of the humors. According to the theory of humors, things hot and dry were superior to those that were cold and moist, because heat was considered the immortal substance of life. Things hot and dry signified masculinity, the sun being one example, and those cold and moist signified femininity, the moon providing the corresponding example. The heat and dryness of men indicated the perfection of their bodies, and women's coldness and moisture indicated their imperfection. Thus the inferiority of woman was determined and defined by her lesser heat, which also explained the interior location of the female genital organs, since women did not possess sufficient heat to propel them outward.

Given the governing logic of this system, it should come as no surprise that even physiological functions that are today exclusively assigned to women's reproductive roles, in particular lactation and menstruation, were at the time conceptualized, to quote Thomas Laqueur, as "part of a common economy of fluids" that was shared by men and women alike and did not entirely define womanhood.[43] Physicians described menstrual blood as the womb's way of providing nourishment for the fetus during pregnancy. In the absence of pregnancy, it was the body's way of expelling the excess blood caused by a surplus of nourishment that women could not otherwise discharge with regularity because they lacked heat and had small veins and a weak digestive system. Therefore, an important function of the menses was to expel cold and decayed blood so as to restore a woman's humoral balance by drying out her body, so to speak. This was a physiological process that protected her from many illnesses.

[42] Shams al-Din al-'Itaqi, *The Treatise on Anatomy of Human Body and Interpretation of Philosophers* [*sic*], Turkish text with English translation and introduction by Esin Kahya (Islamabad: National Hijra Council, 1990), 166. See also al-'Itaqi's description of the uterus in ibid., 116.

[43] Laqueur, *Making Sex*, viii. See also Barbara Duden's exquisite *The Woman Beneath the Skin: A Doctor's Patients in Eighteenth-Century Germany*, translated from German by Thomas Dunlap (Cambridge, MA: Harvard University Press, 1991).

What distinguished women from men was the predictable, cyclical discharge of blood, not the discharge of blood per se. Men, too, expelled blood. Hemorrhoids, for example, were perceived as another way that both the male and female bodies released excess blood and residues that might otherwise collect in body cavities such as the womb, the buttocks, the nose, and the ears, possibly causing hemorrhages. Indeed, one of the posited causes of the stoppage of the menses was the diversion of excess blood to hemorrhoids. This unhealthy stoppage, to which physicians devoted many pages of their treatises, was thought to cause a number of illnesses, ranging from headaches to epileptic fits.[44] In the context of Galenic medicine, then, it is easier to grasp why men with breasts, women with beards, and khunthas were not implausible beings.

The Cultural Landscape of Sexual Categories, Gender Identities, and Gender Roles

Not only were khunthas not implausible beings, but there was also little anxiety about the possibility of their engagement in illicit same-sex acts, an anxiety that pervaded legal and medical writings of hermaphrodites in, for example, Renaissance France.[45] To understand why, it is necessary to venture into the domain of sexual categories, gender identities, and gender roles in the Ottoman Arab world. An exhaustive analysis of this complex subject lies outside the purview of this chapter, but a succinct overview is nonetheless essential at this juncture.

First, it is crucial to appreciate that any attempt to examine same-sex desires and practices through the prism of the modern category of homosexuality produces anachronistic and erroneous readings, for the simple reason that the category of homosexual, with its contemporary meaning of sexual preference as a pivot of identity, simply did not exist in the Middle East (or, for that matter, anywhere else in the world, before the nineteenth century). As Khaled El-Rouayheb's superb monograph on homosexual behavior, titled *Before Homosexuality in the Arab-Islamic World, 1500–1800*, demonstrates, there existed instead

44 Al-Antaki, *Tadhkirat Uli al-Albab*, 2:41–42, 2:140–143; Ibn Sallum, *Ghayat al-Itqan*, 134a, 136a; al-ʻItaqi, *Tashrih al-Abdan*, 117–119.

45 Ibid., 8–9.

a multiplicity of male–male sexual practices and desires that defy location beneath today's category of "homosexuality."[46] I specify "male–male" because we know almost nothing about sex acts between women. One of the rare exceptions I have encountered is a scornful allusion to a sixteenth-century tribade (*musāḥiqa*) with the enigmatic nickname "the Byzantine's daughter."[47] This silence may be due to the fact that nearly all of the extant primary sources were composed by men, and men had little access to the homosocial world of women, where female–female love, desire, and sex would have most likely taken place. Furthermore, men may not have been interested in these acts and desires, as they were not perceived as a serious threat to men's privileged status and to the social order.

Fortunately, Ottoman-era Arabic legal, literary, historical, and medical sources do contain lively and sometimes fiery debates about male same-sex practices. These sources allow us to identify at least four distinct groups of men that might fall under today's category of "homosexual:" the penetrative partner in a sex act (*lūṭī*); the beardless boy (*amrad*); the adult penetrated partner (*ma'būn*); and the "effeminate" (*mukhannath*).

The *Lūṭī*

The penetrative partner in a male–male sexual encounter was usually called *lūṭī*, that is, "one of the people of Lot." In the religious and legal sources, the *lūṭī* (pl. *lāṭā*) was a man who committed a specific act, *liwāṭ* (anal intercourse). Unlike other sexual acts between males, such as kissing, fondling, or intercrural intercourse, anal intercourse was a crime subject to legal sanctions. It was difficult to prove under the law because it required the testimony of eyewitnesses. In addition, the law did not judge sodomy to have occurred unless the participants were all adults. Muhammad al-Tumurtashi (d. 1596), the Hanafi mufti of Gaza, was once asked the following question: "About a boy who committed sodomy with another boy. Is any discretionary punishment

[46] Khaled El-Rouayheb, *Before Homosexuality in the Arab-Islamic World, 1500–1800* (Chicago: Chicago University Press, 2005).

[47] Ibn Ayyub, *al-Rawd al-'Atir*, 72. Samar Habib has recent published *Female Homosexuality in the Middle East: Histories and Representations* (New York: Routledge, 2007) and *Arabo-Islamic Texts on Female Homosexuality, 850–1750 A.D.* (Youngstown, NY: Teneo Press, 2009), which, their titles notwithstanding, are not historical works.

(*taʿzīr*) to be imposed on the two of them?" The mufti replied that neither boy was liable to discretionary punishment, as both were minors.[48]

When the term *lūṭī* appears in nonjuridical sources it is invariably used as an insult to describe a morally dissolute person. To cite but one example, Muhammad al-Saffarini (d. 1744), a Hanbali scholar from Nablus, composed an invective against the sodomites of his time who, he bemoaned, were increasingly present in his homeland. They were often Turks, he asserted, and were recognizable by certain distinctive physical attributes, such as clean-shaven faces and long mustaches, and by specific habits, like frequent congregation in cafes. Al-Saffarini did not mince his words: the sodomites were a plague that had to be suppressed.[49]

Yet there is abundant evidence that, far from being a suppressed practice, sodomy was actually widespread. Even well-respected judges and muftis did it, usually with boys. In sum, the *lūṭī* was viewed as a profligate, but one with both a (arguably hyper) male gender identity and a male gender role that did not preclude marriage or female sexual partners.[50]

The *Amrad*

The objects of both chaste and sexual amorous male attention were usually prepubescent and adolescent boys. These boys, defined as beardless or downy-cheeked (sing. *amrad*), ranged in age from their early teens (sometimes even younger) to as old as twenty. The *amrad*, also referred to as *ghulām* and *ḥadath*, is omnipresent in the Arabic literature of the Ottoman period. As El-Rouayheb has remarked, "much if not most of the extant love poetry of the period is pederastic in tone, portraying an adult male poet's passionate love for a teenage boy."[51]

48 Al-Tumurtashi, *Fatawa*, 35a–35b.

49 Shams al-Din Muhammad al-Saffarini, *Qarʿ al-Siyat fi Qamʿ Ahl al-Liwat*, MS, Chester Beatty Library, Dublin, Ar. 4907(2), 10b–11a. Other authors also composed treatises that strongly condemned *liwāṭ*. See, for instance, *Risala fi Tahrim al-Liwat*, by the mufti of Aleppo Ibrahim al-Sunusi (16th c.). Cited in Ibn al-Hanbali, *Durr al-Habab*, 1:92.

50 For a thorough analysis of sodomy and sodomites, see chapter 3 of El-Rouayheb, *Before Homosexuality*.

51 El-Rouayheb, *Before Homosexuality*, 1. See also El-Rouayheb's exhaustive examination of the *amrad*'s depiction in poetry in "The Love of Boys in Arabic Poetry of the Early Ottoman Period, 1500–1800," *Middle Eastern Literatures* 8, no. 1 (2005):

One explanation for this widespread phenomenon is that children inhabited a "gender limbo" of sorts until they reached full physical development. For boys, that milestone was marked and frequently publicly celebrated by the appearance of the beard, typically at around fifteen.[52] Islamic law decreed that, in the absence of clear physical signs of maturity or of the youth's own declaration of physical maturity, fifteen lunar years was the age at which a boy reached legal majority and became a fully responsible adult.[53] It is unlikely to be a coincidence that fifteen is exactly the age when the girls-turned-boys whose stories I narrated earlier were said to have sprouted a beard and a penis.

There was of course the expectation that children would grow to be men and women, roles for which they were being groomed and trained. Prior to becoming adults, however, their gender remained somewhat ambiguous and fluid. Hence, being the object of male affection, love, or even sexual desire was understandable and was not deemed a threat to a boy's impending masculinity (but would have been fatal to the honor of a girl and of her family), so long as the boundaries of the platonic were not trespassed. Importantly, this "limbo status" of children helps to explain why the genital ambiguity of khunthas was not overly problematic until they came of age.

The *Ma'būn*

Sometimes male–male sex also took place between adult men, in particular between a *lūṭī* or sodomite and a *ma'būn*. A *ma'būn* was the adult male penetrated partner in an illegal sexual act, or, more precisely, a man who desired to be penetrated by other men. Unlike the sodomite, whose "problem" was considered behavioral, the *ma'būn* was usually pathologized. Drawing on and elaborating on the findings of earlier authors, especially the famous medieval physician Abu Bakr

3–22; and Walter Andrews and Mehmet Kalpaklı, *The Age of Beloveds: Love and the Beloved in Early-Modern Ottoman and European Culture and Society* (Durham, NC: Duke University Press, 2005).

52 The famous Damascene mufti and biographer Muhammad Khalil al-Muradi (d. 1791), for example, received poems from several notables when, at about age fifteen, beard-down appeared on his cheeks. Al-Muradi cherished these poems and collected them in a short work titled *Majani al-Thimar min Tahani al-'Idhar*, MS, Chester Beatty Library, Dublin, Ar. 4756(4).

53 Joseph Schacht, *An Introduction to Islamic Law* (Oxford: Clarendon Press, 1964), 124–125.

al-Razi (d. ca. 925), some Ottoman medical treatises stated that a man who desires to be anally penetrated suffered from a medical condition called *ubna*, which could be hereditary. Da'ud al-Antaki, for example, subscribed to the view of *ubna* as hereditary pathology. He attributed it to an irritating substance made of the mineral boron that when present in the veins of the rectum causes the anus to itch and burn, so that relief is sought through anal penetration. Al-Antaki ascribed distinctive physical traits to the *ma'būn*, including a flaccid and fleshy appearance and large buttocks. He recommended a variety of potions and pastes as remedies.[54]

The *Mukhannath*

If representations of sodomite men, beardless boys, and the *ma'būn* are relatively straightforward, that of the *mukhannath* or "effeminate man" is a more elusive category. Khaled El-Rouayheb has proposed that the terms *ma'būn* and *mukhannath* were synonymous and offers substantial evidence that the two words were often used interchangeably.[55]

Some works of physiognomy, the art or science of discerning a person's character from her or his external appearance, corroborate his argument. 'Umar al-Salaqani (fl. 17th c.) posited a direct link between physical signs of effeminacy and the desire for anal penetration. In a tract on physiognomy that in his own estimation was "the most eminent piece of writing composed in this field," al-Salaqani declared that a dullness in a man's eyes accompanied by "a white or black light" in one of them "points to his being effeminate (*isti'nāthihi*) and to his inclination to have intercourse in his rear just as women incline to intercourse in their vaginas." He added "this has been regarded by physicians as a disease" because it results in the transference of a man's sexual desire "from his penis to his rear, may God forbid it!"[56]

An eighteenth-century Egyptian collection of jokes and anecdotes also portrays the *mukhannath* as a man who permits another man to

[54] Al-Antaki, *Tadhkirat Uli al-Albab*, 1:16; al-Antaki, *al-Nuzha al-Mubhija*, 3:105–106; al-Qalyubi, *al-Tadhkira fi al-Tibb*, 79–80. See Franz Rosenthal, "Ar-Razi on the Hidden Illness," *Bulletin of the History of Medicine* 52 (1978): 45–60.

[55] El-Rouayheb, *Before Homosexuality*, 21–22, 45–48.

[56] 'Umar al-Salaqani al-Makki al-Husayni, *Kashf al-Bayan fi Firasat al-Insan*, MS, Chester Beatty Library, Dublin, Ar. 5466, 4a, 10b.

penetrate him during sexual intercourse. According to a joke, "two good-looking effeminates complained about one another to a ruler." The ruler said: "Fuck them," then quickly amended his order to "Beat them!" At that point, one of the effeminates exclaimed: "My lord, your mercy preceded your punishment, so don't take back what you said first!"[57]

These examples suggest that the *ma'būn* and the *mukhannath* were indeed often conflated. On closer analysis, however, it becomes manifest that there were at least two different coexisting attitudes toward effeminates. Like *khunthā*, the word *mukhannath* derives from the Arabic root *kha-na-tha*, which means "to be soft, effeminate." The law made a fine distinction between two types of effeminates. In the eleventh century, a prominent Hanafi jurist argued that one type "behaved corruptly" (presumably the type of effeminate sometimes equated with the *ma'būn*), but the other type was described as "he who has softness in his limbs, talks with the inflection of women, due to his innate constitution, and does not desire women." Some scholars, he added, ruled that it was permissible for this latter type, the one who "does not behave corruptly," to be alone with unrelated women.[58] This crucial distinction was endorsed by later jurists. During a discussion of the categories of persons whose testimony in legal proceedings was acceptable, 'Ala' al-Din al-Haskafi, too, posited two types of effeminates: the *mukhannath*, who "commits bad acts and is penetrated," and the *mukhannith*, who is effeminate "in limbs and speech" "by natural disposition." Because of his corrupt morals, the testimony of the former was not acceptable. The testimony of the latter, on the other hand, was acceptable, presumably because being innately effeminate was not deemed sufficient indication of illicit sexual activity.[59]

This grammatical differentiation between *mukhannath* and *mukhannith* may have been peculiar to certain authors and is rarely encountered outside of the law. It does, however, speak to a conceptual

[57] Al-Bishari, *Bughyat al-Jalis*, 185b, 186a.

[58] Muhammad b. Ahmad al-Sarakhsi (d. ca. 1090), *Kitab al-Mabsut* (Beirut: Dar al-Kutub al-'Ilmiyya, 1993), 5:158. See also Everett K. Rowson, "The Effeminates of Early Medina," *Journal of the American Oriental Society* 111, no. 4 (1991): 675–676.

[59] Al-Haskafi, *al-Durr al-Mukhtar*, 8:198.

difference between two categories of effeminates, those who engage in same-sex sexual acts as the penetrated partner (the *ma'būn*) and those who are effeminate in speech and demeanor but do not "commit bad acts." One *mukhannath* named Qasim in sixteenth-century Aleppo, for example, is described without contempt or ridicule as innately feminine in speech. The appearance of his body was "soft," and he was known for running errands in the streets of Aleppo wearing elevated wooden clogs, a type of shoe reserved for women.[60] The effeminate was thus a distinct category of person, one whose social and physical bodies were in a state of contradiction.

I propose that at least some individuals believed to be *mukhannath* were in fact people with intersex conditions. To mention but three relatively common conditions, Klinefelter syndrome and partial androgen insensitivity syndrome affect the production of male hormones that can result in an effeminate appearance in men, while congenital adrenal hyperplasia causes some females to have a more masculine appearance. This raises the intriguing possibility that the multiplicity, complexity, and fluidity of sex and gender categories in the Ottoman Arab world might have been due, if only in part, to the reality on the ground of ambiguously sexed bodies engendered by the relatively high rate of disorders of sex development in the region, especially hereditary ones. Perhaps nature played a bigger role in shaping attitudes toward sex and gender than strict socioconstruction approaches allow for.

For the time being, in the absence of extensive research, such speculation remains just that. What is clear, however, is that if khunthas are examined in the context of this wide range of gender types and sex acts, it becomes easier to comprehend why adult male–male sex occupied a prominent place in the hierarchy of concerns in the Ottoman Arab East, but khunthas prompted no comparable anxieties.

Conclusion

Atypical or ambiguous genitals were regarded as physical defects in Islamic law, and if they prevented sexual intercourse they could be a serious disability in matters of marriage and reproduction. Khunthas, on the other hand, were placed in a special category of individuals and

60 Ibn al-Hanbali, *Durr al-Habab*, 2:44–45.

jurists dedicated separate chapters of their law books to them. Being labeled khuntha impacted every aspect of a person's life, since every aspect of a person's life was dependent on his or her gender.

When cases of sex indeterminacy came to the attention of the authorities, or when the authorities were asked to intervene, the available tools of the legal and sometimes the medical sciences were mobilized in a vigorous attempt to determine the sex of the khuntha. Practical considerations (inheritance, for example) trumped all others during these attempts to efface sexual indeterminacy. As far as the law was concerned, the objective of these undertakings was the assignment of a definitive gender, the ascription of a social body with which the individual in question was expected to identify and of a gender role that he/she would be required to maintain. Any lingering ambiguity in the physical body was a secondary concern. One's "real" sex (as this concept would be understood in the modern period) was of far less significance.

When the ambiguities were such that assigning a sex was impossible, the khuntha was ruled "ambiguous" (*mushkil*). At least theoretically, the khuntha mushkil was a legal persona endowed with particular, precisely defined, legally codified rights and responsibilities. I employ the qualifying phrase "at least theoretically" because I have not found a single example of an actual khuntha mushkil. That is likely because concerted efforts were made to avoid affixing the label of khuntha mushkil to an adult, as such a designation would have precluded that person's integration into society. So disabling was the state of khuntha mushkil, and so undesirable not only for the individual but also for the family and the community at large, that it was to be avoided if at all possible. Jurists tended to opt for the solutions that caused the least social disruption and gave wide latitude to the khuntha in making the determination, provided that s/he declared his or her sex on reaching physical maturity.

Much like deafness, blindness, and most impairments of the mind, the type of physical difference embodied by khunthas was handled in a frank and straightforward manner, reflecting an attitude generally devoid of embarrassment, shame, or moral opprobrium. There was no insinuation that intersex individuals were the products of divine wrath, and associations of the khuntha with physical, moral, or other forms of corruption were anything but prevalent in the early modern Arab-Islamic world. There is also little evidence that khunthas were ostracized, persecuted, or imprisoned. A notable exception is Muhammad b. Salama's wife, but she was persecuted because she was accused of being an impostor, not because she was a khuntha.

On a similar note, although the accounts found in biographical dictionaries and chronicles display an interest in khunthas as marvels or curiosities (in contrast to the juridical discussions), these narratives rarely, if ever, ascribe freakish or monstrous qualities to people with intersex conditions. Moreover, khunthas were not the objects of sexual anxiety and the possibility of their engagement in unlawful same-sex acts was not a primary concern. In spite of the fact that anxieties about same-sex practices, especially among males, certainly *were* current, khunthas are largely absent from those discussions.

There are multiple and overlapping explanations for these attitudes to intersex and for the relatively high tolerance of sex ambiguity. There was, for one, Islamic law's pragmatism and spirit of accommodation, provided that social stability and status quo were preserved. Second, Islamic theology did not encourage the association of khunthas with sin or deficient spirituality. Third, no rigidly demarcated, impermeable boundaries between the sexes existed in the early modern Arab-Islamic world. Conceptualizations of sex difference were firmly grounded in the notion that the bodies of men and women were variations of a single kind, with the female body constituting a defective version of the male. In other words, differences between female and male were deemed to be quantitative, not qualitative, and hence the boundaries between the sexes were not drawn nearly as firmly as a twenty-first-century reader might expect, and could be traversed if exceptional circumstances warranted it. This relative fluidity must also have been informed, at least in small part, by the reality of the relatively high incidence of intersex conditions in the region.

Conclusion

This journey into an unexplored dimension of the history of the Ottoman Arab world has shown that although physical and mental impairments were identified as such, the categories of "disability" and "disabled" as abstractions and as determinants of identity are anachronistic. To the extent that there was any group identity among the impaired it was probably a localized phenomenon that was confined to the deaf-mute attendants at the sultan's palace in Istanbul and possibly a few other elite households in the Arab lands, or to the blind men at Al-Azhar college for the blind in Cairo, who, at least according to the eighteenth-century Egyptian historian al-Jabarti, had a strong esprit de corps.

People believed that impairments were rooted in the physical body and rarely attributed them to supernatural forces or to the moral, spiritual, or intellectual qualities of the impaired. With the exception of khunthas, for whom no explanation was provided, and holy fools, whose state was ascribed to the ineffable will of the divine, the most commonly cited cause of both physical and mental impairments was the disruption of humoral balance. This explanation is traceable to the Arab-Islamic humoral medical tradition, which drew on and augmented Greco-Roman conceptualizations of the human body, disease etiology, and treatment.

Impairments of the body and the mind rarely resulted in the ascription of social or religious stigma. Arab authors routinely discuss blind and deaf people, people with intersex conditions, and mad people or people with mental impairments in terms that are practical, matter of fact, unapologetic, and largely devoid of anxiety, suspicion, or sensationalism. The reader rarely encounters the positing of a causal relationship between impairments and guilt, sin, ritual impurity, or divine displeasure. Notably, in marked contrast to some contemporaneous European societies, I have found no evidence that the deaf and mute were considered "dumb" or mentally deficient.

This *relatively* benign conceptualization of physical and mental impairments was supported, enabled, and validated by the theory and practice of Islamic law. The jurists' approach consistently displayed the desire to integrate physical minorities into the social body by seeking a balance between the rights and duties of the individual and the interest of the community. For example, the law allowed mute men and women to substitute signs for the written or spoken word when executing legal transactions. In this respect, physical and sensory impairments may be best understood as "situational" disabilities, inasmuch as they were disabling only in particular contexts, like court proceedings in which a blind man could not serve as a witness, or marriage for a woman or a man with certain genital anomalies. Special laws were elaborated for the theoretical khuntha, but in practice most cases of ambiguous sex were transitory and people with intersex conditions either chose a gender for themselves, usually at puberty, or were assigned one by their families or by the authorities.

Impairments of the mind were the most disabling in all spheres of life and were subjected to the most severe legal restrictions, but there was no systematic confinement of "imbeciles" or mad people. Considerable allowances were made for the holy fool, a fluid category that accommodated behaviors ranging from quiet eccentricity to what today might be called psychosis. Regardless of etiology, once an adult was declared to be devoid of reason, her or his legal status was ruled to be that of a minor and a guardian, usually a close relative, was entrusted with all of her or his affairs. Although this arrangement could obviously have disastrous consequences for the person judged insane, depending on the integrity and motivations of the guardian, the law took this action for the protection both of the individual and of the community. Moreover, with the exception of idiocy, which was regarded as a permanent state, impairments of the mind were not necessarily permanent and full rights were regained during periods of lucidity. Ultimately, the state intervened forcefully only when the authorities feared a serious disruption of the status quo.

What lay at the root of these attitudes to impairments? First, if we accept the argument that ignorance of the Other tends to engender fear, and that fear is uppermost in the social production of discrimination and marginalization, then it stands to reason that the presence of large numbers of people with impairments in the Arab lands of the Ottoman Empire would have resulted in reduced discrimination. The relatively high presence of people with impairments was the result of one of the highest rates of consanguineous marriages in the world, in combination with environmental and geological factors such as iodine deficiency in parts of Syria. Thus, early modern Arabs were accustomed to interacting with people with mental and physical impairments (including genital anomalies), and that familiarity may have led to more acceptance and less distrust, suspicion, and marginalization.

Second, religion, as a set of historically grounded and culturally constructed beliefs and practices, undoubtedly played a crucial role in the development of attitudes toward embodied difference that were relatively devoid of stigmatization. As most clearly evidenced by Qur'anic and Hadith pronouncements on blindness, the foundational texts of Islam largely refuse a causal relationship between guilt and disease, illness, and infirmity, and the rejection of the Christian notion of original sin precluded the attribution of congenital defects and impairments to the "sins of the father." In addition, because historically Islamic theology has not placed much emphasis on the devil, the Arab world never witnessed the "wave of scandalous demonic possessions" that swept, for example, sixteenth-century Germany, where Germans' preoccupation with the devil and sin lead to increased persecution of people with certain mental impairments.[1]

A third explanation may be found in the highly stratified nature of early modern society, in which each Muslim individual's socioeconomic status was informed by and enacted through a complex system of constructed norms. It was a world of differences made glaringly visible, one in which a person could be accurately located within the social hierarchy through a glimpse of his or her attire. In the realm of

[1] H. C. Erik Midelfort, *A History of Madness in Sixteenth-Century Germany* (Stanford, CA: Stanford University Press, 1999), 9.

the less immediately visible, geographic origin, ethnicity, and a host of other factors functioned as cardinal constituents of individual and collective identity. In light of this stratification, we can hypothesize that in Arab-Islamic societies, already so complex and featuring such a wealth of Others, people did not seek, designate, or exclude as "ultimate Others" people with physical or mental impairments.

Last but not least is one factor shared by all premodern societies: the reigning mode of production and its structural relationship to labor. Some scholars, including Michael Oliver, Victor Finkelstein, and more recently Lennard Davis, have proposed a materialist interpretation according to which it was only with the rise and expansion of industrialization and industrial capitalism in nineteenth-century Europe that the word "disability" became popular and was conceptualized as a medical pathology. At that time, people with impairments were excluded from wage labor because they could not work at the pace and in the ways demanded by mechanized factory work. Perceived as unproductive members of society, and hence a burden, they had little choice but to turn to charity. Conversely, as Lennard Davis has observed, in preindustrial societies, "[. . .] disability is not as relevant a category as it is under factory conditions where the interchangeability of standardized workers is paramount. People with visual, auditory, or mobility differences can be incorporated into a preindustrial society. Thus, in some sense, their disabilities are not remarkable."[2] Although it merits further investigation, this explanation may also hold true in the preindustrial economy of the early modern Ottoman Arab world, where occupational opportunities were in fact available to impaired individuals, especially blind men.

[2] Lennard Davis, "Dr. Johnson, Amelia, and the Discourse of Disability in the Eighteenth Century," in *"Defects:" Engendering the Modern Body*, ed. Helen Deutsch and Felicity Nussbaum (Ann Arbor: The University of Michigan Press, 2000), 57. See also Michael Oliver, *The Politics of Disablement: A Sociological Approach* (New York: St. Martin's Press, 1990) and Victor Finkelstein, *Attitudes and Disabled People: Issues for Discussion* (New York: World Rehabilitation Fund, 1980).

Epilogue

In 1910, an anonymous article about Najm al-Din al-Ghazzi's famous biographical dictionary of the notables of the sixteenth century was published in the Damascene monthly *al-Muqtabas*, a prominent periodical self-described as "scientific and sociological."[1] The article is essentially a stern condemnation of al-Ghazzi's work, which is therein characterized as the product of an era of Islamic cultural and intellectual decline. This decline, in the author's estimation, was demonstrated by al-Ghazzi's inclusion of "trivial" information in his dictionary: the biographies of holy fools. The writer deemed it shameful that these men should appear on the same pages with the luminaries of the sixteenth century. Those "imbecile fools," he decried, should not even be counted as members of the general populace, "because the populace is superior to them in both reason and faith." "Upon my life!" the author exclaimed, "in a book about the biographies of a century's notables, what is the purpose of inserting among them worthless individuals who trespassed the boundaries of Islamic law with their pretensions that violated mores, and who manipulated the minds of the populace [...]?"[2]

1 "Al-Kawakib al-Sa'ira," *al-Muqtabas: Majalla 'Ilmiyya Ijtima'iyya* 5, no. 12 (1910): 737–749. The article appeared anonymously, but Mahmud al-Shaykh, the editor of al-Ghazzi's *Lutf al-Samar*, suggests that it might have been penned by Muhammad Kurd 'Ali, the editor and owner of *al-Muqtabas*. See the Introduction to *Lutf al-Samar*, 1:78–82.

2 "Al-Kawakib al-Sa'ira," 740.

Muhammad Kurd 'Ali (1876–1953), the editor of *al-Muqtabas* and probably the author of the abovementioned words, was a Muslim secularist reformer, one of the most prominent Syrian exponents of the "Arab Awakening," the cultural and literary revival that began in the late nineteenth century, and one of the founding fathers of Arab journalism.[3] His outrage against those who revered holy fools was by no means a voice in the wilderness, for by the early twentieth century other writers were echoing his sentiments, and the Arabic press featured fierce attacks against the *majādhīb*. In 1929, for example, a piece on holy fools appeared in the Egyptian periodical *al-Manar*, defined on its cover page as "an Islamic magazine that seeks religious, civic, and political reform in all matters." *al-Manar* was edited by Muhammad Rashid Rida (1865–1935), another influential turn-of-the-century Muslim reformer born near Tripoli, in present-day Lebanon. The piece in question is titled "Is the *majdhūb* a saint or a madman?" It consists of a fatwa that Rida issued in response to a reader who sought clarification of the distinction between the holy fool and the insane. Rida condemns the custom of worshipping *majādhīb* in no uncertain terms. They are anything but saints, he asserts, but rather unfortunate individuals afflicted with congenital mental defects, or, much worse, imposters. Their personal appearance and habits – primarily the lack of personal hygiene, the practice of public nudity, and the use of foul or incomprehensible language – more likely indicate, if anything, a relationship with Satan rather than any with the Divine.[4] Now stripped of holiness, holy fools were either recast as clinically ill or were demonized as cunning imposters.

These twentieth-century protestations are surprising because, as this book has shown, only a century earlier members of the same urban, educated strata of society to which Rida and Kurd 'Ali belonged had not hesitated to express openly their belief in the authenticity of many holy fools. Until more research is completed, it would be futile, and perhaps a little arrogant, to attempt to offer any conclusive explanations

[3] On Kurd 'Ali and *al-Muqtabas*, see Samir M. Seikaly, "Damascene Intellectual Life in the Opening Years of the 20th Century: Muhammad Kurd 'Ali and *al-Muqtabas*," in Marwan R. Buheiry, ed., *Intellectual Life in the Arab East, 1890–1939* (Beirut: Center for Arab and Middle East Studies, American University of Beirut, 1981), 125–153.

[4] Muhammad Rashid Rida, "Hal al-Majdhub Wali aw Majnun?," *al-Manar* 30, no. 3 (1929): 190–191.

for how such a shift in attitudes to embodied difference took place, which impairments it benefitted or disadvantaged, and why. It is safe to say, however, that it must have been due to the confluence of multiple factors. Like much of the rest of the world, the Arab lands experienced momentous changes in the nineteenth and twentieth centuries. These included colonialism, the Ottoman Islamic reform movement, Ottoman experiments with political, economic, and legal reform, the introduction of modern means of transportation and communication, attempts at industrialization, intense Christian missionary activity, new philanthropic associations, new forms of state charity, and the proliferation of European-style schools and medical institutions.[5] For example, in the wake of an epidemic of ophthalmia in 1867 that claimed the sight of at least 2,000 people, a School for the Blind was established by the British Syrian Mission in Lebanon and Syria. Its purpose was to instruct the blind in the Bible while simultaneously teaching them handicraft skills. The model of Lebanon's School for the Blind was later followed by other missionary schools for "cripples" as well as for the deaf that opened throughout the region. The Cairo Lunatic Asylum, or 'Abbasiyya, came under the control of the British in 1884, while the Lebanon Hospital for Mental and Nervous Disorders, known locally as 'Asfuriyya, was founded by Swiss Quaker missionaries in 1898 in the outskirts of Beirut with the approval and cooperation of the Ottoman authorities.[6]

The dissemination of the new European sciences of psychiatry and biomedicine was facilitated by these psychiatric (and other) hospitals and by the publication of Arabic-language scientific journals in the late nineteenth century whose principal goal was the popularization of modern Western science.[7] Parts of Charles Darwin's *On the Origin of*

5 Although it does not discuss disability, Mine Ener's *Managing Egypt's Poor and the Politics of Benevolence, 1800–1952* (Princeton, NJ: Princeton University Press, 2003) is an excellent starting point for exploring how new ways of managing charity on the part of the state may have led to the conceptualization of the disabled as a new discrete social category.

6 See Eugene Rogan, "Madness and Marginality: The Advent of the Psychiatric Asylum in Egypt and Lebanon," in *Outside In: On the Margins of the Modern Middle East*, ed. Eugene Rogan (London; New York: I.B. Tauris, 2002), 104–125.

7 On scientific periodicals in Arabic, see Marwa Elshakry, "Darwin's Legacy in the Arab East: Science, Religion and Politics, 1870–1914," Ph.D. diss., Princeton University, 2003, especially chapter 1: "Scientific Missionaries: al-muqtataf and Darwin's Descent into Syria," 19–87.

Species were published in Arabic in 1918, well after the translation of texts of Social Darwinism like the work of Herbert Spencer. Although the philosophy's most extreme form, eugenics, never shaped public policy in the Arab world the way it did in neighboring Kemalist Turkey in the 1930s,[8] it is plausible that the Arab proponents of Social Darwinism contributed to the devaluation of people with congenital impairments (cognitive impairments, congenital blindness and deafness, certain physical deformities), who were viewed as an impediment to progress and the development of a strong and "healthy" nation.

In Egypt, perhaps an even more significant milestone in the formation of new notions of disability was Muhammad ʻAli's (d. 1849) creation of a modern army based on conscription in the first half of the nineteenth century. Egyptian peasants, who formed the majority of this new conscript army, were now divided into fit and unfit, able and disabled, by a controlling state that set the parameters of those categories. Well aware of these parameters, in order to evade the draft some men resorted to self-maiming by, for example, blinding themselves or cutting off their index fingers.[9] This may well have been one of the earliest key moment in the modern history of disability in the Arab East.

[8] Ayça Alemdaroğlu, "Eugenics, Modernity, and Nationalism," in *Social Histories of Disability and Deformity*, ed. David M. Turner and Kevin Stagg (London; New York: Routledge, 2006), 126–141.

[9] Khaled Fahmy, *All the Pasha's Men: Mehmed Ali, His Army and the Making of Modern Egypt* (Cambridge: Cambridge University Press, 1997), 101–102, 260–263.

Bibliography

Unpublished Primary Sources

ʿAfif, Sadaqa (fl. 1729). *Jamiʿ al-Ghara'ib wa-Diwan al-ʿAja'ib*. MS, Staatsbibliothek zu Berlin, We. 1193.

Al-Bishari, Shihab al-Din Ahmad al-Hifnawi (fl. 1769). *Bughyat al-Jalis wa-l-Musamir wa-Nuzhat al-Arwah wa-l-Khawatir*. MS, Chester Beatty Library, Dublin, Ar. 4853.

Al-Burini, Hasan (d. 1615). *Tarajim al-Aʿyan min Abna' al-Zaman*. MS, Chester Beatty Library, Dublin, Ar. 3219 (1).

Al-Hanuti, Muhammad b. ʿUmar (d. 1601/2). *Muntakhab al-Fatawa al-Hanuti* [*sic*]. MS, Chester Beatty Library, Dublin, Ar. 5280.

Al-Husayni, Hasan (d. 1811). *Fatawa*. MS, Princeton University Library, Yahuda 5415.

Ibn Budayr, Shihab al-Din Ahmad al-Hallaq (fl. 1762). *Hawadith Dimashq al-Sham al-Yawmiyya min Sanat 1154 ila Sanat 1176*. MS, Chester Beatty Library, Dublin, Ar. 3551(2).

Ibn al-Ha'ik, Ismaʿil (d. 1701/2). *Shifa' al-ʿAlil bi-Fatawa al-Marhum al-Shaykh Ismaʿil*. MS, Princeton University Library, Yahuda 5535.

Ibn Sallum, Salih b. Nasr Allah al-Halabi (d. 1670/1). *Ghayat al-Itqan fi Tadbir Badan al-Insan*. MS, Dar al-Kutub al-Wataniyya, Cairo, al-Madrasa al-Ahmadiyya 1044.

Al-Makki, Jar Allah Muhammad b. Fahd (d. 1547). *Al-Nukat al-Ziraf fi al-Mawʿiza bi-Dhawi al-ʿAhat min al-Ashraf*. MS, Chester Beatty Library, Dublin, Ar. 3838.

______ *Mühimme-i Mısır*, vol. 6, no. 601 (late Jumada I, 1163/April or May 1750). Başbakanlık Osmanlı Arşivi, Istanbul.

Al-Muradi, Muhammad Khalil (d. 1791). *Majani al-Thimar min Tahani al-ʿIdhar*. MS, Chester Beatty Library, Dublin, Ar. 4756(4).

Sabuncuoğlu, Şerefeddin (d. ca. 1468). *Cerrahiyyetü'l Haniyye*. MS, Bibliothèque nationale de France, Département des manuscrits, Supplément turc 693.

Al-Saffarini, Shams al-Din Muhammad (d. 1744). *Qar' al-Siyat fi Qam' Ahl al-Liwat*. MS, Chester Beatty Library, Dublin, Ar. 4907(2).

Al-Salaqani, 'Umar b. Khalil al-Makki al-Husayni (fl. 17th c.). *Kashf al-Bayan fi Firasat al-Insan*. MS, Chester Beatty Library, Dublin, Ar. 5466.

Al-Sharabati, 'Abd al-Karim (d. 1764). *Al-Thabat al-Mubarak*. MS, Chester Beatty Library, Dublin, Ar. 4273.

Al-Tarabulusi, 'Abd Allah al-Khalili (fl. 18th c). *Fatawa*. MS, Princeton University Library, Yahuda 507.

Al-Tumurtashi, Muhammad b. 'Abd Allah al-Ghazzi (d. 1596). *Fatawa*. MS, Princeton University Library, Yahuda 4135.

Published Primary Sources

Ali, Ahmed. *Al-Qur'an: A Contemporary Translation*. Princeton, NJ: Princeton University Press, 1988.

Alpin, Prosper (d. 1617). *La Médecine des Egyptiens par Prosper Alpin, 1581–1584*. Edited and translated from Latin by R. de Fenoyl. 2 vols. Cairo: Institut Français d'Archéologie Orientale, 1980.

[Anonymous]. "Al-Kawakib al-Sa'ira." *Al-Muqtabas: Majalla 'Ilmiyya Ijtima'iyya* 5, no. 12 (1910): 737–749.

Al-Antaki, Da'ud (d. 1599). *Tadhkirat Uli al-Albab wa-l-Jami' li-l-'Ajab al-'Ujab*. Beirut: Al-Maktaba al-Thaqafiyya, n.d.

——— *Al-Nuzha al-Mubhija fi Tashhidh al-Adhhan wa-Ta'dil al-Amzija*. Printed on the margins of al-Antaki's *Tadhkirat Uli al-Albab* [see previous entry].

Al-Baydawi, 'Abd Allah (d. ca. 1386). *Anwar al-Tanzil wa-Asrar al-Ta'wil*. 2 vols. Beirut: Dar al-Kutub al-'Ilmiyya, 1999.

Biddulph, William (d. 17th c.). "Part of a Letter of Master William Biddulph from Aleppo." In Samuel Purchas, *Hakluytus Posthumus, or, Purchas His Pilgrimes: Contayning a History of the World in Sea Voyages and Lande Travells by Englishmen and Others*, 8:248–304. Glasgow: James MacLehose and Sons, 1905.

Bon, Ottaviano (d. 1623). *The Sultan's Seraglio: An Intimate Portrait of Life at the Ottoman Court*. Introduced and annotated by Godfrey Goodwin. London: Saqi Books, 1996.

Al-Budayri (see Ibn Budayr above), Ahmad Shihab al-Din (fl. 1762). *Hawadith Dimashq al-Yawmiyya 1154–1175/1741–1762*. Edited by Ahmad 'Izzat 'Abd al-Karim. N.p.: Matba'at Lajnat al-Bayan al-'Arabi, 1959. [This edition was based on the 19th-century recension of the original text by the Damascene scholar Muhammad Sa'id al-Qasimi].

Al-Bukhari, Muhammad b. Isma'il (d. 870). *[Sahih al-Bukhari]: The Translation of the Meanings of Sahih Al-Bukhari: Arabic-English*. Translated by Muhammad Muhsin Khan. 2nd rev. ed. 9 vols. Medina, Saudi Arabia: Islamic University, 1973–1976.

Al-Burini, Hasan (d. 1615). *Tarajim al-A'yan min Abna' al-Zaman*. Edited by Salah al-Din al-Munajjid. 2 vols. Damascus: Matbu'at al-Majma' al-'Ilmi al-'Arabi bi-Dimashq, 1959 and 1963. Part 3 was never published, but is available in manuscript form. See "Unpublishd Primary Sources."

Çelebi, Evliya (d. 1682). *Al-Rihla ila Misr wa-l-Sudan wa-l-Habasha*. Edited and translated from Turkish into Arabic by Muhammad Harb. 2 vols. Madinat Nasr, Cairo: Dar al-Afaq al-'Arabiyya, 2006.

______ *Evliya Tshelebi's Travels in Palestine (1648–1650)*. Translated from Turkish by St. H. Stephan. Jerusalem: Ariel Publishing House, 1980.

Dallam, Thomas (d. after 1614). *Diary for 1599: Account of an Organ Carryed to the Grand Seignor and Other Curious Matter. Early Voyages and Travels in the Levant. I – The Diary of Master Thomas Dallam, 1599–1600*. Edited by J. Theodore Bent. London: Hakluyt Society, 1893.

Al-Dasuqi, Muhammad b. Ahmad b. 'Arafah (d. 1815). *Hashiyat al-Dasuqi 'ala al-Sharh al-Kabir*. 2 vols. Cairo: Dar Ihya' al-Kutub al-'Arabiyya, n.d.

Erizzo, Antonio. "Sommario della relazione di Antonio Erizzo bailo a Costantinopoli, letta in Senato nel 1557." In *Relazioni degli ambasciatori veneti al Senato durante il secolo decimosesto*, edited by Eugenio Albèri, vol. 9 (series 3, vol. 3). Firenze: Società editrice fiorentina, 1855.

Al-Ghazzi, Najm al-Din Muhammad (d. 1651). *Al-Kawakib al-Sa'ira bi-A'yan al-Mi'a al-'Ashira*. Edited by Jibra'il Sulayman Jabbur. 2nd edition. 3 vols. Beirut: Manshurat Dar al-Afaq al-Jadida, 1979.

______ *Lutf al-Samar wa-Qatf al-Thamar min Tarajim A'yan al-Tabaqa al-Ula min al-Qarn al-Hadi 'Ashar*. Edited by Mahmud al-Shaykh. 2 vols. Damascus: Manshurat Wizarat al-Thaqafa wa-l-Irshad al-Qawmi, 1981–1982.

Al-Halabi, Ibrahim b. Muhammad (d. 1549/50). *Multaqa al-Abhur*. Edited by Wahbi Sulayman Ghawji al-Albani. Beirut: Mu'assasat al-Risala, 1989.

Al-Hamawi, Ahmad b. Muhammad (d. 1686/7). *Ghamz 'Uyun al-Basa'ir Sharh Kitab al-Ashbah wa-l-Naza'ir*. 4 vols. Beirut: Dar al-Kutub al-'Ilmiyya, 1985. This is al-Hamawi's commentary of Zayn al-'Abidin Ibn Nujaym's (d. 1563) famous Hanafi tract *al-Ashbah wa-l-Naza'ir*.

Al-Haskafi, 'Ala' al-Din Muhammad (d. 1677). *Al-Durr al-Mukhtar Sharh Tanwir al-Absar*. In Ibn 'Abidin, *Radd al-Muhtar* [see Ibn 'Abidin].

______ *Al-Durr al-Muntaqa fi Sharh al-Multaqa*. Printed on the margins of *Majma' al-Anhur fi Sharh Multaqa al-Abhur* by 'Abd al-Rahman Shaykhzadah. 4 vols. in 2. Beirut: Dar al-Kutub al-'Ilmiyya, 1998.

Ibn 'Abidin, Muhammad Amin (d. 1836). *Radd al-Muhtar 'ala al-Durr al-Mukhtar Sharh Tanwir al-Absar*. Edited by 'Adil Ahmad 'Abd al-Mawjud and 'Ali Muhammad Mu'awwad. 2nd edition. 10 vols. Beirut: Dar al-Kutub al-'Ilmiyya, 2003.

______ *Al-'Uqud al-Durriyya fi Tanqih al-Fatawa al-Hamidiyya*. Cairo: Al-Matba'a al-'Amira al-Miriyya, 1882/3.

Ibn Ayyub al-Ansari, Sharaf al-Din Musa (d. ca. 1592). *Nuzhat al-Khatir wa-Bahjat al-Nazir*. Edited by 'Adnan Muhammad Ibrahim. 2 vols. Damascus: Manshurat Wizarat al-Thaqafa, 1991.

_____ *Al-Rawd al-'Atir fi-ma Tayassara min Akhbar Ahl al-Qarn al-Sabi' ila Khitam al-Qarn al-'Ashir*. Partial edition by Ahmet Halil Güneş with the title *Das Kitāb ar-raud al-'ātir des Ibn Aiyūb: Damaszener Biographien des 10./16. Jahrhunderts, Beschreibung und Edition*. Berlin: Klaus Schwarz Verlag, 1981.

Ibn al-Hanbali, Radi al-Din Muhammad (d. 1563). *Bahr al-'Awamm fi-ma Asaba fi-hi al-'Awamm*. Edited by Sha'ban Salah. Cairo: Dar al-Thaqafa al-'Arabiyya, 1990.

_____ *Durr al-Habab fi Tarikh A'yan Halab*. Edited by Mahmud Ahmad al-Fakhuri and Yahya Zakariya 'Abbara. 4 vols. Damascus: Manshurat Wizarat al-Thaqafa, 1972–1974.

Ibn al-Himsi, Shihab al-Din Ahmad b. Muhammad (d. 1527/8). *Hawadith al-Zaman wa-Wafayat al-Shuyukh wa-l-Aqran*. 3 vols. Edited by 'Umar 'Abd al-Salam al-Tadmuri. Sidon; Beirut: al-Maktaba al-'Asriyya, 1999.

Ibn al-'Imad, 'Abd al-Hayy b. Ahmad (d. 1679). *Shadharat al-Dhahab fi Akhbar Man Dhahab*. 8 vols. in 4. Beirut: Al-Maktab al-Tijari li-l-Tiba'a wa-l-Nashr wa-l-Tawzi', 1966.

Ibn Jum'a, Muhammad (fl. 1743). *Al-Bashat wa-l-Qudah*. Edited by Salah al-Din al-Munajjid in *Wulat Dimashq fi al-'Asr al-'Uthmani*. Damascus: n.p., 1949.

Ibn Kannan al-Salihi, Muhammad (d. 1740). *Al-Hawadith al-Yawmiyya min Tarikh Ihda 'Ashar wa-Alf wa-Mi'a*. Partial edition by Akram Hasan al-'Ulabi with the title *Yawmiyyat Shamiyya*. Damascus: Dar al-Tabba', 1994.

Ibn Kathir, Isma'il (d. 1373). *Tafsir Ibn Kathir (Abridged)*. Abridged by a group of scholars under the supervision of Safi-ur-Rahman al-Mubarakpuri. 10 vols. Riyadh: Darussalam, 2000.

Ibn Nujaym, Zayn al-'Abidin (d. 1563). *Al-Ashbah wa-l-Naza'ir*. Edited with a commentary by Ahmad b. Muhammad al-Hamaqi (d. 1686/7) in *Ghamz 'Uyun al-Basa'ir Sharh Kitab al-Ashbah wa-l-Naza'ir*. 4 vols. Beirut: Dar al-Kutub al-'Ilmiyya, 1985.

Ibn Sallum, Salih b. Nasr Allah al-Halabi (d. 1670/1). Book Four of *Ghayat al-Itqan fi Tadbir Badan al-Insan* was edited by Kamal Chehadeh in *al-Tibb al-Jadid al-Kimiya'i*. Aleppo: Manshurat Jami'at Halab, Ma'had al-Turath al-'Ilmi al-'Arabi, 1997.

Ibn Tulun, Shams al-Din Muhammad (d. 1546). *Mufakahat al-Khillan fi Hawadith al-Zaman: Tarikh Misr wa-l-Sham*. Edited by Muhammad Mustafa. 2 vols. Cairo: Al-Mu'assasa al-Misriyya al-'Amma li-l-Ta'lif wa-l-Tarjama, 1962–1964.

Al-'Imadi, Hamid b. 'Ali (d. 1758). [*Fatawa*] As edited by Ibn 'Abidin in *al-'Uqud al-Durriyya fi Tanqih al-Fatawa al-Hamidiyya*. Cairo: Al-Matba'a al-'Amira al-Miriyya, 1882/3.

Al-'Itaqi, Shams al-Din (fl. 1632). *The Treatise on Anatomy of Human Body and Interpretation of Philosophers* [*sic*]. Turkish text with English translation and introduction by Esin Kahya. Islamabad: National Hijra Council, 1990.

Al-Jabarti, 'Abd al-Rahman (d. 1825/6). *'Aja'ib al-Athar fi al-Tarajim wa-l-Akhbar*. Edited by Hasan Muhammad Jawhar, 'Umar al-Dasuqi, and al-Sayyid Ibrahim Salim. 7 vols. Cairo: Lajnat al-Bayan al-'Arabi, 1959–1967.

_____ *'Abd al-Rahman al-Jabarti's History of Egypt: 'Aja'ib al-Athar fi'l-Tarajim wa'l-Akhbar*. 5 vols. in 3. Edited by Thomas Philipp and Moshe Perlmann. Stuttgart, Germany: Franz Steiner Verlag, 1994.

Al-Jahiz, Abu 'Uthman (d. 868/9). *Kitab al-Bursan wa-l-'Umyan wa-l-Hulan*. Edited by Muhammad Mursi al-Khuli. Beirut: Mu'assasat al-Risala, 1981.

Al-Kawakibi, Muhammad b. Hasan (d. 1685). *Al-Fawa'id al-Samiyya fi Sharh al-Fara'id al-Saniyya*. 2 vols. Bulaq: Al-Matba'a al-Kubra al-'Amiriyya, 1904/05.

Leiser, Gary, and Michael Dols. "Evliyā Chelebi's Description of Medicine in Seventeenth-Century Egypt. Part II: Text." *Sudhoffs Archiv* 72, no. 1 (1988): 49–68.

Al-Marghinani, Burhan al-Din 'Ali b. Abi Bakr (d. 1196/7). *Al-Hidaya: Sharh Bidayat al-Mubtadi*. 4 vols. Edited by Muhammad Muhammad Tamir and Hafiz 'Ashur Hafiz. Cairo: Dar al-Salam, 2000.

_____ *The Hedàya, or Guide; A Commentary on the Mussulman Laws: Translated by the Order of the Governor-General and Council of Bengal by Charles Hamilton*. 4 vols. London: Printed by T. Bensley, 1791.

Al-Muhibbi, Muhammad Amin (d. 1699). *Khulasat al-Athar fi A'yan al-Qarn al-Hadi 'Ashar*. 4 vols. Beirut: Dar Sadir, 1966. Reprint of the 1867/8 Cairo edition.

Al-Muradi, Muhammad Khalil (d. 1791). *'Arf al-Basham fi-man Waliya Fatwa Dimashq al-Sham*. Edited by Muhammad Muti' al-Hafiz and Riyad 'Abd al-Hamid Murad. 2nd edition. Damascus; Beirut: Dar Ibn Kathir, 1988.

_____ *Silk al-Durar fi A'yan al-Qarn al-Thani 'Ashar*. Edited by Akram Hasan al-'Ulabi. 4 vols. Beirut: Dar Sadir, 2001.

Al-Nabulusi, 'Abd al-Ghani (d. 1731). *Al-Hadra al-Unsiyya fi al-Rihla al-Qudsiyya*. Edited by Akram Hasan al-'Ulabi. Beirut: Al-Masadir, 1990.

_____ *Ta'tir al-Anam fi Ta'bir al-Manam*. 2 vols. 2nd edition. Cairo: Dar Ihya' al-Kutub al-'Arabiyya, 1958–1959.

Al-Nawawi, Yahya b. Sharaf (d. 1277). *Adab al-Fatwa wa-l-Mufti wa-l-Mustafti*. Edited by Bassam 'Abd al-Wahhab Jabi. Damascus: Dar al-Fikr, 1988.

Al-Qalyubi, Ahmad (d. 1659). *Al-Tadhkira fi al-Tibb*. Edited by Ahmad Farid al-Mazidi and Muna Shalabi with the title *Tadhkirat al-Qalyubi fi al-Tibb wa-l-Hikma*. Beirut: Dar al-Kutub al-'Ilmiyya, 2001.

Al-Qari, 'Ali b. Sultan Muhammad al-Harawi (d. 1606). *Tasliyat al-A'ma 'ala Baliyat al-'Ama*. Edited by Mabruk Isma'il Mabruk. Cairo: Maktabat al-Qur'an, 1994.

Al-Qurtubi, Muhammad (d. 1273). *Al-Jami' li-Ahkam al-Qur'an*. 10 vols. Cairo: Dar al-Kitab al-'Arabi li-l-Tiba'a wa-l-Nashr, 1967.

Al-Ramli, Khayr al-Din (d. 1671). *Al-Fatawa al-Khayriyya li-Naf' al-Barriyya.* 2 vols in 1. Cairo: n.p., 1857.

Rida, Muhammad Rashid. "Hal al-Majdhub Wali aw Majnun?" *Al-Manar* 30, no. 3 (1929): 190–191.

Russell, Alexander (d. 1768). *The Natural History of Aleppo. Containing a Description of the City, and its Principal Natural Productions in its Neighborhood. Together with an Account of the Climate, Inhabitants, and Diseases; Particularly on the Plague.* 2 vols. 2nd edition, revised, enlarged, and illustrated with notes by Patrick Russell. Hants., London: Gregg International, 1969 (reprint of: London: Printed for G. G. and J. Robinson, 1794).

Rycaut, Paul (d. 1700). *The History of the Present State of the Ottoman Empire.* New York: Arno Press, 1971. First published in 1668.

Al-Safadi, Khalil b. Aybak (d. 1363). *Nakt al-Himyan fi Nukat al-'Umyan.* Edited by Tariq Tantawi. Cairo: Maktabat al-Tala'i, 1997.

——— *Al-Shu'ur bi-l-'Ur.* Edited by 'Abd al-Razzaq Husayn. Amman: Dar 'Ammar, 1988.

Al-Sarakhsi, Muhammad b. Ahmad (d. ca. 1090). *Kitab al-Mabsut.* 15 vols. Beirut: Dar al-Kutub al-'Ilmiyya, 1993.

Al-Sha'rani, 'Abd al-Wahhab (d. 1565). *Lawahiq al-Anwar fi Tabaqat al-Akhyar.* Cairo: Maktabat al-Adab, 1993. Also known as *al-Tabaqat al-Kubra.* Only vol. 1 has been published thus far.

——— *Lawahiq al-Anwar fi Tabaqat al-Akhyar*, also known as *al-Tabaqat al-Kubra.* Abridged translation by Virginia Vacca in *Vite e detti di santi musulmani.* Milano: TEA, 1988. Originally published in Torino by UTET in 1968.

——— *Lata'if al-Minan wa-l-Akhlaq fi Wujub al-Tahadduth bi-Ni'mat Allah 'ala al-Itlaq.* 2 vols. Cairo: n.p., 1938.

Slade, Adolphus (d. 1877). *Records of Travels in Turkey, Greece, etc., and of a Cruise in the Black Sea with the Capitan Pasha in the Years 1829, 1830 and 1831.* Philadelphia and Baltimore: Carey and Hart, 1833.

Spink, M. S., and G. L. Lewis, eds. and trans. *Albucasis on Surgery and Instruments: A Definitive Edition of the Arabic Text with English Translation and Commentary.* Berkeley and Los Angeles: University of California Press, 1973.

Al-Talawi, Darwish (d. 1606). *Sanihat Duma al-Qasr fi Mutarahat Bani al-'Asr.* Edited by Muhammad Musa al-Khawli. 2 vols. Beirut: 'Alam al-Kutub, 1983.

Tietze, Andreas. *Mustafa 'Ali's Description of Cairo of 1599: Text, Transliteration, Translation, Notes.* Vienna: Verlag der Österreichischen Akademie der Wissenschaften, 1975.

Al-Tumurtashi, Muhammad b. 'Abd Allah al-Ghazzi (d. 1596). *Matn Tanwir al-Absar wa-Jami' al-Bihar.* Cairo: Matba'at al-Taraqqi, 1914.

Al-'Urdi, Abu al-Wafa' b. 'Umar al-Halabi (d. 1660). *Ma'adin al-Dhahab fi al-A'yan al-Musharrafa bi-him Halab.* Edited by Muhammad Altunji. Damascus: Dar al-Mallah, 1987.

——— *Ma'adin al-Dhahab fi al-A'yan al-Musharrafa bi-him Halab*. Edited by 'Abd Allah al-Ghazali. Al-Safah, Kuwait: Maktabat Dar al-'Uruba, 1987.

——— *Ma'adin al-Dhahab fi al-A'yan al-Musharrafa bi-him Halab*. Edited by 'Isa Sulayman Abu Salim. Amman: Manshurat Markaz al-Watha'iq wa-l-Makhtutat, al-Jami'a al-Urdunniyya, 1992.

Al-Zabidi, Muhammad Murtada (d. 1791). *Taj al-'Arus*. 10 vols. Benghazi, Libya: Dar Libya li-l-Nashr wa-l-Tawzi', 1966.

Bey, Ali. *Travels of Ali Bey in Morocco, Tripoli, Cyprus, Egypt, Arabia, Syria, and Turkey. Between the Years 1803 and 1807. Written by Himself, and Illustrated with Maps and Numerous Plates*. Second American from the First London Edition. In Two Volumes. Philadelphia: Printed for John Conrad, at the Shakespeare Buildings, 1816.

Hathaway, Jane, ed. *Al-Jabarti's History of Egypt*. Princeton, NJ: Markus Wiener Publishers, 2009.

Secondary Literature

'Abbud, Khazin. *Al-Khulafa' wa-l-Fuqaha' wa-l-Shu'ara' wa-l-Udaba' al-'Umyan*. Beirut: Dar al-Harf al-'Arabi, 2004.

'Abd Allah, Layla. *Ahkam al-Akhras fi al-Fiqh al-Islami*. Cairo: Maktabat al-Zahra' li-l-Tab' wa-l-Nashr wa-l-Tawzi', 1997.

'Abd al-Razzaq, Husayn. *Al-I'aqa fi al-Adab al-'Arabi*. Al-Shariqa, UAE: Manshurat Madinat al-Shariqa li-l-Khadamat al-Insaniyya, 1999.

Abd El-Khalek, Amira. "The Cultural Construction of Women with Disabilities in Egypt: An Ethnographic Approach." *The MIT Electronic Journal of Middle East Studies* 4 (Fall 2004): 92–107.

Abdel-Fattah, M. A. "Arabic Sign Language: A Perspective." *Journal of Deaf Studies and Deaf Education* 10, no. 2 (2005): 212–221.

Abdel Haleem, M. A. S. "The Blind and the Qur'an." *Journal of Qur'anic Studies* 3, no. 2 (2001): 123–125.

Abdullah, M. A., et al. "Ambiguous Genitalia: Medical, Socio-Cultural and Religious Factors Affecting Management in Saudi Arabia." *Annals of Tropical Paediatrics* 11, no. 4 (1991): 343–348.

Abrams, Judith Z. *Judaism and Disability: Portrayals in Ancient Texts from the Tanach Through the Bavli*. Washington, DC: Gallaudet University Press, 1998.

Abu-Habib, Lina, ed. *Gender and Disability: Women's Experiences in the Middle East*. Oxford: Oxfam, 1997.

Abu Habib, Sa'di. *Al-Mu'awwaq wa-l-Mujtama' fi al-Shari'a al-Islamiyya*. Damascus: Dar al-Fikr, 1982.

Adivar, Abdülhak Adnan. *La science chez les Turcs ottomans*. Paris: G.-P. Maisonneuve, 1939.

Afifi, H. H., et al. "The Most Encountered Groups of Genetic Disorders in Giza Governorate, Egypt." *Bratislava Medical Journal* 111, no. 2 (2010): 62–69.

Akesode, Ademola F., Walter J. Meyer III, and Claude J. Migeon. "Male Pseudohermaphroditism with Gynaecomastia Due to Testicular17-Ketosteroid Reductase Deficiency." *Clinical Endocrinology* 7, no. 6 (Dec. 1977): 443–452.

Akkach, Samer. *'Abd al-Ghani al-Nabulusi: Islam and Enlightenment.* Oxford: Oneworld, 2007.

Alaani, Samira et al., "Pilot Study of Congenital Anomaly Rates at Birth in Fallujah, Iraq, 2010." *Journal of the Islamic Association of North America* 44 (2012): 1–7.

Al-Abdulwahab, Sami S., and Salah I. Al-Gain. "Attitudes of Saudi Arabian Health Care Professionals Towards People with Physical Disabilities." *Asia Pacific Disability Rehabilitation Journal* 14, no. 1 (2003): 63–70.

Alemdaroğlu, Ayça. "Eugenics, Modernity, and Nationalism." In *Social Histories of Disability and Deformity*, edited by David M. Turner and Kevin Stagg, 126–141. London; New York: Routledge, 2006.

Al-Fityani, Kinda. "Arab Sign Languages: A Lexical Comparison." *Center for Research in Language Technical Reports* 19, no. 1 (2007): 3–13.

Al-Fityani, Kinda, and Carol Padden. "A Lexical Comparison of Sign Languages in the Arab World." In *Sign Languages: Spinning and Unraveling the Past, Present and Future. TISLR9, Forty Five Papers and Three Posters from the 9th Theoretical Issues in Sign Language Research Conference, Florianopolis, Brazil, December 2006*, edited by R. M. de Quadros, 2–14. Petrópolis, Brazil: Editora Arara Azul, 2008.

Al-Issa, Ihsan, ed. *Al-Junūn: Mental Illness in the Islamic World.* Madison, CT: International University Press, 2000.

_____ "Mental Illness in Medieval Islamic Society." In *Al-Junūn: Mental Illness in the Islamic World*, edited by Ihsan Al-Issa, 43–69. Madison, CT: International University Press, 2000.

Al-Maghribi, Hussein. "Congenital Adrenal Hyperplasia: Problems with Developmental Anomalies of the External Genitalia and Sex Assignment." *Saudi Journal of Kidney Diseases and Transplantation* 18, no. 3 (2007): 405–413.

Amar, Zohar. *The History of Medicine in Jerusalem.* Oxford: Archaeopress, 2002.

Andrews, Walter, and Mehmet Kalpaklı. *The Age of Beloveds: Love and the Beloved in Early-Modern Ottoman and European Culture and Society.* Durham, NC: Duke University Press, 2005.

'Anuti, Usama. *Al-Haraka al-Adabiyya fi Bilad al-Sham Khilal al-Qarn al-Thamin 'Ashar.* Beirut: Al-Jami'a al-Lubnaniyya, 1970.

Arabi, Oussama. "The Regimentation of the Subject: Madness in Islamic and Modern Arab Civil Laws." In *Standing Trial: Law and the Person in the Modern Middle East*, edited by Baudouin Dupret, 264–293. London; New York: I.B. Tauris, 2004.

Arnaout, M. A. "Late-Onset Congenital Adrenal Hyperplasia in Women with Hirsutism." *European Journal of Clinical Investigation* 22, no. 10 (1992): 651–658.

Auchterlonie, Paul. *Arabic Biographical Dictionaries: A Summary Guide and Bibliography*. Durham, UK: Middle East Libraries Committee, 1987.

Barasch, Moshe. *Blindness: The History of a Mental Image in Western Thought*. New York: Routledge, 2001.

Barnes, Colin, and Geof Mercer. *Exploring Disability*. 2nd edition. Malden, MA: Polity Press, 2010.

Basha, ʿUmar Musa. *Tarikh al-Adab al-ʿArabi: Al-ʿAsr al-ʿUthmani*. Beirut: Dar al-Fikr al-Muʿasir; Damascus: Dar al-Fikr, 1999.

Baynton, Douglas C. "Disability and the Justification of Inequality in American History." In *The New Disability History: American Perspectives*, edited by Paul K. Longmore and Lauri Umansky, 33–57. New York: New York University Press, 2001.

Bazna, Maysaa S., and Tarek A. Hatab. "Disability in the Qur'an: The Islamic Alternative to Defining, Viewing, and Relating to Disability." *Journal of Religion, Disability and Health* 9, no. 1 (2005): 5–27.

Ben-Moshe, Liat, and Sami Colligan. "The State of Disability in Israel/Palestine: An Introduction." *DSQ: Disability Studies Quarterly* 27, no. 4 (2007): n.p.

Benzahra, Saloua Ali. "Representations of the Disabled in Arab/Islamic Culture and Literature from North Africa and the Middle East." PhD diss., The University of Minnesota, 2002.

Berger, Lutz. *Gesellschaft und Individuum in Damaskus, 1550–1791*. Würzburg, Germany: Ergon, 2007.

Blackless, Melanie, et al. "How Sexually Dimorphic Are We? Review and Synthesis." *American Journal of Human Biology* 12, no. 2 (2000): 151–166.

Bonner, Michael, Mine Ener, and Amy Singer, eds. *Poverty and Charity in Middle Eastern Contexts*. Albany: State University of New York Press, 2003.

Bouhdiba, Abdelwahab. *Sexuality in Islam*. Translated from French by Alan Sheridan. London: Saqi Books, 1998. Originally published in 1975.

Branson, Jan, and Don Miller. *Damned for Their Difference: The Cultural Construction of Deaf People as Disabled*. Washington, DC: Gallaudet University Press, 2002.

Bray, Julia. "Dawud ibn ʿUmar al-Antaki." In *Essays in Arabic Literary Biography, 1350–1850*, edited by Joseph E. Lowry and Devin J. Stewart, 48–56. Wiesbaden, Germany: Harrassowitz Verlag, 2009.

Brisson, Luc. *Sexual Ambivalence: Androgyny and Hermaphroditism in Graeco-Roman Antiquity*. Translated from French by Janet Lloyd. Berkeley: University of California Press, 2002.

Burch, Susan, and Ian Sutherland. "Who's Not Yet Here? American Disability History." *Radical History Review* 94 (Winter 2006): 127–147.

Busby, Chris, Malak Hamdan, and Entesar Ariabi. "Cancer, Infant Mortality and Birth-Sex Ratio in Fallujah, Iraq, 2005–2009." *International Journal of Environmental Research and Public Health* 7 (2010): 2828–2837.

Bynum, Caroline. "Why All the Fuss about the Body? A Medievalist's Perspective." *Critical Inquiry* 22 (Autumn 1995): 1–33.

Cadden, Joan. *Meanings of Sex Difference in the Middle Ages: Medicine, Science, and Culture*. Cambridge: Cambridge University Press, 1993.

Campbell, Fiona Kumari. *Contours of Ableism: The Production of Disability and Abledness*. New York: Palgrave Macmillan, 2009.

Cilardo, Agostino. "Historical Development of the Legal Doctrine Relative to the Position of the Hermaphrodite in the Islamic Law." *The Search: Journal for Arab and Islamic Studies* 7 (1986): 128–170.

Cleminson, Richard, and Francisco Vásquez García. *Hermaphroditism, Medical Science and Sexual Identity in Spain, 1850–1960*. Cardiff, UK: University of Wales Press, 2010.

Cockayne, Emily. "Experiences of the Deaf in Early Modern England." *The Historical Journal* 46, no. 3 (2003): 495–510.

Cockburn, Patrick. "Toxic Legacy of U.S. Assault in Fallujah 'Worse Than Hiroshima'." *The Independent*, July 24, 2010. Available online at http://www.independent.co.uk/news/world/middle-east/toxic-legacy-of-us-assault-on-fallujah-worse-than-hiroshima-2034065.html. Accessed July 6, 2013.

Conrad, Lawrence I. "Arab-Islamic Medicine." In *Companion Encyclopedia of the History of Medicine*, edited by W. F. Bynum and Roy Porter, 1: 676–727. New York: Routledge, 1993.

______ "Islam and the Body." In *The Oxford Companion to the Body*, edited by Colin Blakemore and Sheila Jennett, 392–394. Oxford; New York: Oxford University Press, 2001.

Dankoff, Robert. *An Ottoman Mentality: The World of Evliya Çelebi*. Leiden; Boston: Brill, 2004.

Daston, Lorraine, and Katharine Park. "Hermaphrodites in Renaissance France." *Critical Matrix* 1, no. 5 (1985): 1–19.

______ "The Hermaphrodite and the Orders of Nature: Sexual Ambiguity in Early Modern France." In *Premodern Sexualities*, edited by Louise Fradenburg and Carla Freccero, 117–136. New York; London: Routledge, 1996.

Davis, Lennard J. *Enforcing Normalcy: Disability, Deafness, and the Body*. London; New York: Verso, 1995.

______, ed. *The Disability Studies Reader*. New York: Routledge, 1997.

______ *Bending over Backwards: Disability, Dismodernism, and Other Difficult Positions*. New York: New York University Press, 2002a.

______ "Dr. Johnson, Amelia, and the Discourse of Disability in the Eighteenth Century." In *"Defects:" Engendering the Modern Body*, edited by Helen Deutsch and Felicity Nussbaum, 54–74. Ann Arbor: The University of Michigan Press, 2000b.

Delague, Valérie, et al. "Mutational Analysis in Lebanese Patients with Congenital Adrenal Hyperplasia Due to a Deficit in 21-Hydroxylase." *Hormone Research* 53, no. 2 (2000): 77–82.

Dessouky, Nabil M. "Gender Assignment for Children with Intersex Problems: An Egyptian Perspective." *Egyptian Journal of Surgery* 20, no. 2 (2001): 499–515.

Deutsch, Helen, and Felicity Nussbaum, eds. *"Defects": Engendering the Modern Body*. Ann Arbor: The University of Michigan Press, 2000.

Diamond, Milton. "Pediatric Management of Ambiguous and Traumatized Genitalia." *The Journal of Urology* 162, no. 3 (September 1999): 1021–1028.

Dikici, Ayşe Ezgi. "Imperfect Bodies, Perfect Companions? Dwarfs and Mutes at the Ottoman Court in the Sixteenth and Seventeenth Centuries." Master's thesis, Sabancı University, 2006.

Dols, Michael W. *The Black Death in the Middle East*. Princeton, NJ: Princeton University Press, 1977.

______ "Leprosy in Medieval Arabic Medicine." *Journal of the History of Medicine and Allied Sciences* 34, no. 3 (1979): 314–333.

______ "The Leper in Medieval Islamic Society." *Speculum* 58, no. 4 (1983): 891–916.

______ *Medieval Islamic Medicine: Ibn Ridwan's Treatise "On the Prevention of Bodily Ills in Egypt."* Berkeley: University of California Press, 1984.

______ "Insanity and Its Treatment in Islamic Society." *Medical History*: 31, no. 1 (1987): 1–14.

______ *Majnūn: The Madman in Medieval Islamic Society*. Edited by Diana E. Immisch. Oxford: Clarendon Press, 1992.

Dordain, Claude Marie. *Les membres non résidents étrangers de l'Académie Royale de Chirurgie*. Thèse de Rennes no. 397 (1964).

Douglas, Mary. *Purity and Danger: An Analysis of Concepts of Pollution and Taboo*. New York: Praeger, 1966.

Dreger, Alice Domurat. *Hermaphrodites and the Medical Invention of Sex*. Cambridge, MA: Harvard University Press, 1998.

Duden, Barbara. *The Woman Beneath the Skin: A Doctor's Patients in Eighteenth-Century Germany*. Translated from German by Thomas Dunlap. Cambridge, MA: Harvard University Press, 1991.

Eckstein, Benjamin, Shulamit Cohen, Amicur Farkas, and Ariel Rösler. "The Nature of the Defect in Familial Male Pseudohermaphroditism in Arabs of Gaza." *The Journal of Clinical Endocrinology and Metabolism* 68, no. 2 (1989): 477–485.

Edwards, Martha L. "Constructions of Physical Disability in the Ancient Greek World: The Community Concept." In *The Body and Physical Difference: Discourses of Disability*, edited by David T. Mitchell and Sharon L. Snyder, 35–50. Ann Arbor: The University of Michigan Press, 1997.

Eiesland, Nancy L. *The Disabled God: Toward a Liberatory Theology of Disability*. Nashville, TN: Abingdon Press, 1994.

Eiesland, Nancy L., and Don E. Saliers, eds. *Human Disability and the Service of God: Reassessing Religious Practice*. Nashville, TN: Abingdon Press, 1998.

El-Rouayheb, Khaled. *Before Homosexuality in the Arab-Islamic World, 1500–1800*. Chicago: University of Chicago Press, 2005a.

______ "The Love of Boys in Arabic Poetry of the Early Ottoman Period, 1500–1800." *Middle Eastern Literatures* 8, no. 1 (2005b): 3–22.

Elshakry, Marwa. "Darwin's Legacy in the Arab East: Science, Religion and Politics, 1870–1914." PhD diss., Princeton University, 2003.

El-Zawahreh, Taisir Khalil Muhammad. *Religious Endowments and Social Life in the Ottoman Province of Damascus in the Sixteenth and Seventeenth Centuries*. Karak, Jordan: Mu'tah University, 1992.

Ener, Mine. *Managing Egypt's Poor and the Politics of Benevolence, 1800–1952*. Princeton, NJ: Princeton University Press, 2003.

Epstein, Julia. "Either/Or – Neither/Both: Sexual Ambiguity and the Ideology of Gender." *Genders* 7 (Spring 1990): 99–142.

Ernst, Waltraud, ed. *Histories of the Normal and the Abnormal: Social and Cultural Histories of Norms and Normativity*. London; New York: Routledge, 2006.

Ewing, Katherine Pratt. *Arguing Sainthood: Modernity, Psychoanalysis, and Islam*. Durham, NC: Duke University Press, 1997.

Eyler, Joshua R., ed. *Disability in the Middle Ages: Reconsiderations and Reverberations*. Burlington, VT: Ashgate, 2010.

Fahmy, Khaled. *All the Pasha's Men: Mehmed Ali, His Army and the Making of Modern Egypt*. Cambridge: Cambridge University Press, 1997.

_____ "Women, Medicine, and Power in Nineteenth-Century Egypt." In *Remaking Women: Feminism and Modernity in the Middle East*, edited by Lila Abu Lughod, 34–72. Princeton, NJ: Princeton University Press, 1998.

Fausto-Sterling, Anne. *Sexing the Body: Gender Politics and the Construction of Human Sexuality*. New York: Basic Books, 2000.

Feher, Michel, Ramona Naddaf, and Nadia Tazi, eds. *Fragments for a History of the Human Body*. 3 vols. New York: Zone, 1989.

Finkelstein, Victor. "To Deny or Not to Deny Disability." In *Handicap in a Social World: A Reader*, edited by Ann Brechin, Penny Liddiard, and John Swain, 34–36. Sevenoaks, UK: Hodder and Stoughton, 1981. Originally published in *The Magic Carpet* 27, no. 1 (1975): 31–38.

_____ *Attitudes and Disabled People: Issues for Discussion*. New York: World Rehabilitation Fund, 1980.

Foucault, Michel. *Herculine Barbin, dite Alexina B*. Paris: Gallimard, 1978a.

_____ *The History of Sexuality. Volume 1, An Introduction; Volume 2, The Uses of Pleasure; Volume 3, The Care of the Self*. Translated from French by Robert Hurley. New York: Pantheon, 1978b, 1985, 1986.

_____ *The Birth of the Clinic: An Archaeology of Medical Perception*. Translated from French by A. M. Sheridan Smith. New York: Vintage, 1994.

_____ *History of Madness*. Edited by Jean Khalfa and translated from French by Jonathan Murphy and Jean Khalfa. London; New York: Routledge, 2006.

Garland, Robert. *The Eye of the Beholder: Deformity and Disability in the Graeco-Roman World*. Ithaca, NY: Cornell University Press, 1995.

Geoffrey, Eric. *Le soufisme en Égypte et en Syrie sous les derniers Mamelouks et les premiers Ottomans: orientations spirituelles et enjeux culturels*. Damascus: Institut français de Damas; Paris: diff. A. Maisonneuve, 1995.

Gerber, Haim. *Islamic Law and Culture, 1600–1840*. Leiden; Boston: Brill, 1999.

Ghaly, Mohammed. "Physical and Spiritual Treatment of Disability in Islam: Perspectives of Early and Modern Jurists." *Journal of Religion, Disability and Health* 12, no. 2 (2008): 105–143.

_____ "Physiognomy: A Forgotten Chapter of Disability in Islam. The Discussions of Muslim Jurists." *Bibliotheca Orientalis* 66, nos. 3–4 (2009): 161–197.

_____ *Islam and Disability: Perspectives in Theology and Jurisprudence*. London; New York: Routledge, 2010.

Gharaibeh, Numan. "Disability in Arab Societies." In *Disabilities: Insights from Across Fields and Around the World*, edited by Catherine A. Marshall, Elizabeth Kendall, Martha E. Banks, and Reva Mariah S. Gover, 1: 63–79. Westport, CT: Praeger, 2009.

Gibb, H. A. R. "Islamic Biographical Literature." In *Historians of the Middle East*, edited by Bernard Lewis and P. M. Holt, 54–58. London: Oxford University Press, 1962.

Gilbert, Ruth. "'Strange Notions': Treatments of Early Modern Hermaphrodites." In *Madness, Disability and Social Exclusion: The Archaeology and Anthropology of 'Difference'*, edited by Jane Hubert, 144–158. London; New York: Routledge, 2000.

_____ *Early Modern Hermaphrodites: Sex and Other Stories*. New York: Palgrave, 2002.

Goffman, Erving. *Stigma: Notes on the Management of Spoiled Identity*. Englewood Cliffs, NJ: Prentice-Hall, 1963.

Goodley, Dan. *Disability Studies: An Interdisciplinary Introduction*. Los Angeles, CA; London: SAGE Publications, 2011.

Gracer, Bonnie L. "What the Rabbis Heard: Deafness in the Mishnah." *Disability Studies Quarterly* 23, no. 2 (Spring 2003): 192–205.

Graham, William A. *Beyond the Written Word: Oral Aspects of Scripture in the History of Religion*. New York: Cambridge University Press, 1987.

Greenberg, Julie A. *Intersexuality and the Law: Why Sex Matters*. New York: New York University Press, 2012.

Grehan, James Paul. "The Mysterious Power of Words: Language, Law, and Culture in Ottoman Damascus (17th-18th Centuries)." *Journal of Social History* 37, no. 4 (Summer 2004): 991–1015.

_____ *Everyday Life and Consumer Culture in Eighteenth-Century Damascus*. Seattle: University of Washington Press, 2007.

Groce, Nora Ellen. *Everyone Here Spoke Sign Language: Hereditary Deafness on Martha's Vineyard*. Cambridge, MA: Harvard University Press, 1985.

_____ "The Cultural Context of Disability." In *Genetics, Disability, and Deafness*, edited by John Vickery Van Cleve, 23–30. Washington, DC: Gallaudet University Press, 2004.

Habib, Samar. *Female Homosexuality in the Middle East: Histories and Representations*. New York: Routledge, 2007.

_____ *Arabo-Islamic Texts on Female Homosexuality, 850–1750 A.D.* Youngstown, NY: Teneo Press, 2009.

Al-Hafiz, Muhammad Muti'. *Faqih al-Hanafiyya Muhammad Amin 'Abidin: Hayatuhu wa-Atharuhu*. Beirut: Dar al-Fikr al-Mu'asir; Damascus: Dar al-Fikr, 1994.

Haj, Fareed. *Disability in Antiquity*. New York: Philosophical Society, 1970.

Hamidan, Zuhayr. *A'lam al-Hadara al-'Arabiyya al-Islamiyya fi al-'Ulum al-Asasiyya wa-l-Tatbiqiyya fi al-'Ahd al-'Uthmani*. Damascus: Manshurat Wizarat al-Thaqafa, 1996.

Harper, Catherine. *Intersex*. Oxford; New York: Berg, 2007.

Hathaway, Jane, with Karl K. Barbir. *The Arab Lands under Ottoman Rule, 1516–1800*. Harlow, UK: Pearson Longman, 2008.

Hawwa, Muhammad b. Mahmud. *Huquq Dhawi al-Ihtiyajat al-Khassa fi al-Shari'a al-Islamiyya*. Beirut: Dar Ibn Hazm li-l-Tiba'a wa-l-Nashr wa-l-Tawzi', 2010.

Hendriks, Bernadet. *Jordanian Sign Language: Aspects of Grammar from a Cross-Linguistic Perspective*. Utrecht, The Netherlands: LOT, 2008.

Herdt, Gilbert, ed. *Third Sex, Third Gender: Beyond Sexual Dimorphism in Culture and Society*. New York: Zone Books, 1994.

Heyd, Uriel. *Studies in Old Ottoman Criminal Law*. Edited by V. L. Menage. Oxford: Clarendon Press, 1973.

Hochberg, Ze'ev, et al. "Clinical, Biochemical, and Genetic Findings in a Large Pedigree of Male and Female Patients with 5α-Reductase 2 Deficiency." *Journal of Clinical Endocrinology and Metabolism* 81, no. 8 (August 1996): 2821–2827.

Hoffman, Valerie J. "Islamic Perspectives on the Human Body: Legal, Social and Spiritual Considerations." In *Embodiment, Morality, and Medicine*, edited by Lisa Sowle Cahill and Margaret A. Farley, 37–55. Dordrecht, The Netherlands; Boston: Kluwer Academic Publishers, 1995.

Huard, Pierre. "Castration judiciaire en 1726 à Tripoli (de Syrie)." *Clio Medica* 2 (1967): 363–369.

Imber, Colin. *Ebu's-su'ud: The Islamic Legal Tradition*. Stanford, CA: Stanford University Press, 1997.

Ingstad, Benedicte, and Susan Reynolds Whyte, eds. *Disability and Culture*. Los Angeles: University of California Press, 1995.

Ismail, S. I., and I. Mazen. "A Study of Gender Outcome of Egyptian Patients with 46,XY Disorder of Sex Development." *Sexual Development* 4, nos. 4–5 (2010): 285–291.

Jacquart, Danielle, and Claude Thomasset. *Sexuality and Medicine in the Middle Ages*. Translated from French by Matthew Adamson. Cambridge: Polity Press, 1988.

Jaussen, J. A. "Le Cheikh Sa'ad ad-Din et les Djinn, à Naplouse." *Journal of the Palestine Oriental Society* 1–4 (1920–1924): 145–157.

Jervis, John. *Transgressing the Modern: Explorations in the Western Experience of Otherness*. Oxford; Malden, MA: Blackwell Publishers, 1999.

Johansen, Baber. "The Valorization of the Human Body in Muslim Sunni Law," *Princeton Papers: Interdisciplinary Journal of Middle Eastern Studies* 4 (1996): 71–112.

Jones, Ann Rosalind, and Peter Stallybrass, "Fetishizing Gender: Constructing the Hermaphrodite in Renaissance Europe." In *Body Guards: The Politics of Gender Ambiguity*, edited by Julia Epstein and Kristina Straub, 80–111. New York: Routledge, 1991.

Kahya, Esin, and Ayşegül Demirhan Erdemir. *Medicine in the Ottoman Empire (And Other Scientific Developments)*. Istanbul: Nobel Medical Publications, 1997.

Kaidbey, Naila Takieddine. "Historiography in *Bilad al-Sham*: The Sixteenth and Seventeenth Centuries." PhD diss., American University of Beirut, 1995.

Kandemir, N., and N. Yordam. "Congenital Adrenal Hyperplasia in Turkey: A Review of 273 Patients." *Acta Paediatrica* 86 (1997): 22–25.

Karamustafa, Ahmet. "The Antinomian Dervish as Model Saint." In *Modes de transmission de la culture religieuse en Islam*, edited by Hassan Elboudari, 241–250. Cairo: Institut Français d'Archéologie Orientale, 1993.

_____ *God's Unruly Friends: Dervish Groups in the Islamic Later Middle Period, 1200–1550*. Salt Lake City: University of Utah Press, 1994.

Kashani-Sabet, Firoozeh. "The Haves and the Have Nots: A Historical Study of Disability in Modern Iran." *Iranian Studies* 43, no. 2 (2010): 167–195.

Khalidi, Tarif. "Islamic Biographical Dictionaries: A Preliminary Assessment." *The Muslim World* 63, no. 1 (1973): 53–65.

King, Helen. *Hippocrates' Woman: Reading the Female Body in Ancient Greece*. New York: Routledge, 1998.

Kisch, Shifra. "Negotiating (Genetic) Deafness in a Bedouin Community." In *Genetics, Disability, and Deafness*, edited by John Vickery Van Cleve, 148–173. Washington, DC: Gallaudet University Press, 2004.

_____ "Disablement, Gender, and Deafhood among the Negev Arab-Bedouin." *Disability Studies Quarterly* 27, no. 4 (2007): n.p.

Kohn, G., E. E. Lasch, R. El Shawwa, Y. Litvin, and A. Rosler. "Male Pseudohermaphroditism Due to 17ß-Hydroxysteroid Dehydrogenase Deficiency (17ßHSD) in a Large Arab Kinship. Studies on the Natural History of the Defect." *Journal of Pediatric Endocrinology* 1, no. 1 (1985): 29–37.

Kudlick, Catherine. "Disability History: Why We Need Another 'Other'." *The American Historical Review* 108, no. 3 (June 2003): 763–793.

Kugle, Scott. "Vision and Blindness." In *Encyclopaedia of the Qur'an*, edited by Jane McAuliffe, 5: 445–447. Leiden: Brill, 2005.

_____ *Sufis and Saints' Bodies: Mysticism, Corporeality, and Sacred Power*. Chapel Hill: The University of North Carolina Press, 2007.

Lababidi, Lesley, with Nadia El-Arabi. *Silent No More: Special Needs People in Egypt*. Cairo; New York: The American University in Cairo Press, 2002.

Laes, Christian. "Silent Witnesses: Deaf-Mutes in Graeco-Roman Antiquity." *Classical World* 104, no. 4 (2011): 451–473.

Laqueur, Thomas. *Making Sex: Body and Gender from the Greeks to Freud*. Cambridge, MA: Harvard University Press, 1990.

_____ "Sex in the Flesh." *Isis* 94 (2003): 300–306.

Leclerc, Lucien. *Histoire de la médecine arabe*. 2 vols. Paris: Ernest Leroux, 1876.

Leiser, Gary, and Michael Dols. "Evliyā Chelebi's Description of Medicine in Seventeenth-Century Egypt. Part I: Introduction." *Sudhoffs Archiv* 71, no. 2 (1987): 197–216.

Lev, Yaacov. "The Ethics and Practice of Islamic Medieval Charity." *History Compass* 5, no. 2 (2007): 603–618.

Lewis, Bernard. "Studies in the Ottoman Archives – I." *Bulletin of the School of Oriental and African Studies, University of London* 16, no. 3 (1954): 469–501.

Longmore, Paul K. *Why I Burned My Book and Other Essays on Disability*. Philadelphia: Temple University Press, 2003.

Longmore, Paul K., and Lauri Umansky, eds. *The New Disability History: American Perspectives*. New York: New York University Press, 2001.

MacDonald, Michael. *Mystical Bedlam: Madness, Anxiety and Healing in Seventeenth-Century England*. Cambridge; New York: Cambridge University Press, 1981.

Mahone, Sloan, and Megan Vaughan, eds. *Psychiatry and Empire*. New York: Palgrave Macmillan, 2007.

Malti-Douglas, Fedwa. "Pour une rhétorique onomastique: les noms des aveugles chez as-Safadi." *Cahiers d'Onomastique Arabe* 1 (1979): 7–19.

_____ "Dreams, the Blind, and the Semiotics of the Biographical Notice." *Studia Islamica* 51 (1980): 137–162.

_____ "Mentalités and Marginality: Blindness and Mamlūk Civilization." In *The Islamic World from Classical to Modern Times: Essays in Honor of Bernard Lewis*, edited by Clifford Edmund Bosworth et al., 211–238. Princeton, NJ: Darwin Press, 1989.

Marcus, Abraham. *The Middle East on the Eve of Modernity: Aleppo in the Eighteenth Century*. New York: Columbia University Press, 1993.

Marsot, Afaf Lufti al-Sayyid, ed. *Society and the Sexes in Medieval Islam*. Malibu, CA: Undena, 1979.

Masud, Muhammad Khalid, Brinkley Messick, and David S. Powers, eds. *Islamic Legal Interpretation: Muftis and Their Fatwas*. Cambridge, MA: Harvard University Press, 1996.

Mayers, Marilyn A. "A Century of Psychiatry: The Egyptian Mental Hospitals." PhD diss., Princeton University, 1984.

Mazen I., et al. "Screening of Genital Anomalies in Newborns and Infants in Two Egyptian Governorates." *Hormone Research in Paediatrics* 73, no. 6 (2010): 438–442.

Mazen, I., O. Hiort, R. Bassiouny, and M. El Gammal. "Differential Diagnosis of Disorders of Sex Development in Egypt." *Hormone Research* 70, no. 2 (2008): 118–123.

McDonagh, Patrick. *Idiocy: A Cultural History*. Liverpool: Liverpool University Press, 2009.

McRuer, Robert. *Crip Theory: Cultural Signs of Queerness and Disability*. New York: New York University Press, 2006.

Meekosha, Helen. "Decolonising Disability: Thinking and Acting Globally." *Disability and Society* 26, no. 6 (October 2011): 667–682.

Meekosha, Helen, and Karen Soldatic. "Human Rights and the Global South: The Case of Disability." *Third World Quarterly* 32, no. 8 (2011): 1383–1398.

Metzler, Irina. *Disability in Medieval Europe: Thinking About Physical Impairment During the High Middle Ages, c. 1100–1400*. London: Routledge, 2006.

——— "Perceptions of Deafness in the Central Middle Ages." In *Homo debilis: Behinderte – Kranke – Versehrte in der Gesellschaft des Mittelalters*, edited by Cordula Nolte, 79–98. Korb, Germany: Didymos-Verlag, 2009.

——— "Disability in the Middle Ages: Impairment at the Intersection of Historical Inquiry and Disability Studies." *History Compass* 9, no. 1 (2011): 45–60.

——— *A Social History of Disability in the Middle Ages: Cultural Considerations of Physical Impairment*. New York: Routledge, 2013.

Meyerhof, Max. "The History of Trachoma Treatment in Antiquity and During the Arabic Middle Ages." *Bulletin of the Ophthalmological Society of Egypt* 24 (1936): 26–87.

Meyerhof, Max, and M. Monnerot-Dumaine. "Quelques maladies d'Europe dans une encyclopédie médicale arabe du XVIIe siècle." *Bulletin de l'Institut d'Égypte* 24 (1942): 33–47.

Midelfort, H. C. Erik. *A History of Madness in Sixteenth-Century Germany*. Stanford, CA: Stanford University Press, 1999.

Miles, M. "Disability in an Eastern Religious Context: Historical Perspectives." *Disability and Society* 10, no. 1 (1995): 49–69.

——— "Signing in the Seraglio: Mutes, Dwarfs and Jestures at the Ottoman Court 1500–1700." *Disability and Society* 15, no. 1 (2000): 115–134.

——— "Disability and Deafness in the Middle East: A Bibliography Comprising Materials with Technical, Cultural and Historical Relevance to Child and Adult Impairments, Disabilities and Deafness, Incapacity, Mental Disorders, Special Needs, Social and Educational Responses and Rehabilitation; Partly Annotated." 2008. Available at http://cirrie.buffalo.edu/bibliography/mideast/index.php. Hosted by the Center for International Rehabilitation Research Information and Exchange.

——— "Deaf People, Sign Language and Communication, in Ottoman and Modern Turkey: Observations and Excerpts from 1300 to 2009. From Sources in English, French, German, Greek, Italian, Latin and Turkish, with Introduction and Some Annotation," 2009. Available at http://www.independentliving.org/miles200907.html.

Mirzoeff, Nicholas. "Framed: The Deaf in the Harem." In *Deviant Bodies: Critical Perspectives on Difference in Science and Popular Culture*, edited by Jennifer Terry and Jacqueline Urla, 49–77. Bloomington: Indiana University Press, 1995.

——— *Silent Poetry: Deafness, Sign, and Visual Culture in Modern France*. Princeton, NJ: Princeton University Press, 1995.

Misconi, Humam, and Maryam Navi. "Medical Geology in the Middle East." In *Medical Geology: A Regional Synthesis*, edited by Olle Selinus, Robert B. Finkelman, and Jose A. Centeno, 135–174. Dordrecht, The Netherlands; London: Springer, 2010.

Mitchell, David T., and Sharon L. Snyder, eds. *The Body and Physical Difference: Discourses of Disability*. Ann Arbor: The University of Michigan Press, 1997.

_____ *Narrative Prosthesis: Disability and the Dependencies of Discourse*. Ann Arbor: The University of Michigan Press, 2000.

Mor, Sagit. "'Tell My Sister to Come and Get Me Out of Here:' A Reading of Ableism and Orientalism in Israel's Immigration Policy (The First Decade)." *Disability Studies Quarterly* 27, no. 4 (2007): n.p.

Morad, M., Y. Nasri, and J. Merrick. "Islam and the Person with Intellectual Disability." In *Spirituality and Intellectual Disability*, edited by W. C. Gaventa, and D. L. Coulter, 65–72. New York: The Haworth Pastoral Press, 2001.

Murphey, Rhoads. "Ottoman Medicine and Transculturalism From the Sixteenth Through the Eighteenth Century." *Bulletin of the History of Medicine* 66 (1992): 376–403.

Musallam, Basim. *Sex and Society in Islam: Birth Control before the Nineteenth Century*. Cambridge: Cambridge University Press, 1983.

Nakamura, Karen. *Deaf in Japan: Signing and the Politics of Identity*. Ithaca, NY: Cornell University Press, 2006.

Nederman, Cary J., and Jacqui True. "The Third Sex: The Idea of the Hermaphrodite in Twelfth-Century Europe." *Journal of the History of Sexuality* 6, no. 4 (1996): 497–517.

Nelson, Kristina. *The Art of Reciting the Qur'an*. Austin: University of Texas Press, 1985.

Oliver, Michael. *The Politics of Disablement: A Sociological Approach*. New York: St. Martin's Press, 1990.

Olyan, Saul M. *Disability in the Hebrew Bible: Interpreting Mental and Physical Differences*. New York: Cambridge University Press, 2008.

Park, Katharine, and Robert A. Nye. "Destiny Is Anatomy." *The New Republic* 204 (February 18, 1991): 53–57.

Pedani, Maria Pia. "Il silenzio del sultano." In *Il potere della parola, la parola del potere. Tra Europa e mondo arabo-ottomano, tra Medioevo ed età moderna*, edited by Antonella Ghersetti, 99–112. Venice: Filippi Editore, 2010.

Peters, Rudolph. *Crime and Punishment in Islamic Law: Theory and Practice from the Sixteenth to the Twenty-First Century*. Cambridge; New York: Cambridge University Press, 2005.

Plann, Susan. *A Silent Minority: Deaf Education in Spain, 1550–1835*. Berkeley: University of California Press, 1997.

Porter, Roy. "History of the Body." In *New Perspectives on Historical Writing*, edited by Peter Burke, 206–232. University Park: The Pennsylvania State University Press, 1992.

_____ *Madness: A Brief History*. Oxford; New York: Oxford University Press, 2002.

Pothier, Dianne, and Richard Devlin, eds. *Critical Disability Theory: Essays in Philosophy, Politics, Policy, and Law*. Vancouver: University of British Columbia Press, 2006.

Rafeq, Abdul-Karim. *The Province of Damascus, 1723–1783*. Beirut: Khayats, 1966.

Raz, Aviad E., and Marcela Atar. "Upright Generations of the Future: Tradition and Medicalization in Community Genetics." *Journal of Contemporary Ethnography* 33, no. 3 (2004): 296–322.

Rebstock, Ulrich. "Mathematischen Quellen zur Rechtsgeschichte: Das Problem des Hermaphroditen." *Die Welt des Orients* 20–21 (1990): 99–114.

Reis, Elizabeth. *Bodies in Doubt: An American History of Intersex*. Baltimore, MD: The Johns Hopkins University Press, 2009.

Reynolds, Dwight F., ed. *Interpreting the Self: Autobiography in the Arabic Literary Tradition*. Berkeley: University of California Press, 2001.

Richardson, Kristina L. *Difference and Disability in the Medieval Islamic World: Blighted Bodies*. Edinburgh: Edinburgh University Press, 2012a.

_____ "Disability? Perspectives on Bodily Difference From the Middle East." In *Why the Middle Ages Matter: Medieval Light on Modern Injustice*, edited by Celia Ghazelle Simon Doubleday, Felice Lifshitz, and Amy G. Remensnyder, 121–129. London; New York: Routledge, 2012b.

Rippin, Andrew. "Seeing and Hearing." In *Encyclopaedia of the Qur'an*, edited by Jane McAuliffe, 4: 573–576. Leiden: Brill, 2004.

Rispler-Chaim, Vardit. *Disability in Islamic Law*. Dordrecth, The Netherlands: Springer, 2007.

Roded, Ruth. *Women in Islamic Biographical Collections: From Ibn Sa'd to Who's Who*. Boulder, CO: Lynne Rienner Publishers, 1994.

Rogan, Eugene. "Madness and Marginality: The Advent of the Psychiatric Asylum in Egypt and Lebanon." In *Outside In: On the Margins of the Modern Middle East*, edited by Eugene Rogan, 104–125. London; New York: I. B. Tauris, 2002.

Rose, Martha. *The Staff of Oedipus: Transforming Disability in Ancient Greece*. Ann Arbor: The University of Michigan Press, 2003.

Rosenthal, Franz. "Ar-Razi on the Hidden Illness." *Bulletin of the History of Medicine* 52 (1978): 45–60.

Rösler, Ariel. "Steroid 17ß-Hydroxysteroid Dehydrogenase Deficiency in Man: An Inherited Form of Male Pseudohermaphroditism." *Journal of Steroid Biochemistry and Molecular Biology* 43, no. 8 (December 1992): 989–1002.

Rösler, Ariel, and Gertrude Kohn. "Male Pseudohermaphroditism Due to 17ß-Hydroxysteroid Dehydrogenase Deficiency: Studies on the Natural History of the Defect and Effect of Androgens on Gender Role." *Journal of Steroid Biochemistry* 19, no. 1 (1983): 663–674.

Rowson, Everett K. "The Categorization of Gender and Sexual Irregularity in Medieval Arabic Vice Lists." In *Body Guards: The Cultural Politics of*

Gender Ambiguity, edited by Julia Epstein and Kristina Straub, 50–79. New York: Routlegde, 1991a.

_____ "The Effeminates of Early Medina." *Journal of the American Oriental Society* 111, no. 4 (1991b): 671–693.

_____ "Gender Irregularity as Entertainment: Institutionalized Transvestism at the Caliphal Court in Medieval Baghdad." In *Gender and Difference in the Middle Ages*, edited by Sharon Farmer and Carol Braun Pasternack, 45–72. Minneapolis: University of Minnesota Press, 2003.

Russell, G. A. "Salih ibn Nasr-Allah al-Halabi ibn Sallum." In *Dizionario Biografico della Storia della Medicina e delle Scienze Naturali*, edited by Roy Porter, 4: 76. Florence: Franco Maria Ricci, 1989.

_____ "The Emergence of Physiological Optics." In *Encyclopedia of the History of Arabic Science*, edited by Roshdi Rashed, 672–715. London; New York: Routledge, 1996.

Al-Sabbagh, Layla. *Min A'lam al-Fikr al-'Arabi fi al-'Asr al-'Uthmani al-Awwal: Muhammad al-Amin al-Muhibbi al-Mu'arrikh wa-Kitabuhu Khulasat al-Athar fi A'yan al-Qarn al-Hadi 'Ashar*. Damascus: al-Sharika al-Muttahida li-l-Tawzi', 1986.

Sabra, Adam. *Poverty and Charity in Medieval Islam: Mamluk Egypt, 1250-1517*. Cambridge: Cambridge University Press, 2000.

Sadan, Yusuf. *Al-Adab al-'Arabi al-Hazil wa-Nawadir al-Thuqala': Al-'Ahat wa-l-Masawi' al-Insaniyya wa-Makanatuha fi al-Adab al-Raqi*. Köln, Germany: Al-Kamel Verlag, 2007. Originally published in 1983.

Saidi, Aliya. "Marriage and Mental Illness in the Mamluk Period." In *Towards a Cultural History of the Mamluk Era*, edited by Mahmoud Haddad, Armin Heinemanna, John L. Meloy, and Souad Slim, 153–158. Beirut: Orient-Institut Beirut; Würzburg, Germany: Ergon in Kommission, 2010.

Sajdi, Dana. "Peripheral Visions: The Worlds and Worldviews of Commoner Chroniclers in the 18th-Century Ottoman Levant." PhD diss., Columbia University, 2002.

Sanders, Paula. "Gendering the Ungendered Body: Hermaphrodites in Medieval Islamic Law." In *Women in Middle Eastern History: Shifting Boundaries in Sex and Gender*, edited by Nikki R. Keddie and Beth Baron, 74–95. New Haven, CT: Yale University Press, 1991.

Sarı, Nil. "The Classification of Mental Diseases in the Ottoman Medical Manuscripts," *Tıp Tarihi Araştırmaları* 1 (1986): 105–112.

Sarı, Nil, and M. Bedizel Zülfikar, "The Paracelsusian [*sic*] Influence on Ottoman Medicine in the Seventeenth and Eighteenth Centuries." In *Transfer of Modern Science and Technology to the Muslim World: Proceedings of the International Symposium on "Modern Sciences and the Muslim World" (Istanbul 2–4 September 1987)*, edited by Ekmeleddin Ihsanoğlu, 157–179. Istanbul: Research Centre for Islamic History, Art and Culture, 1992.

Savage-Smith, Emilie. "Drug Therapy of Eye Diseases in Seventeenth-Century Islamic Medicine: The Influence of the 'New Chemistry' of the Paracelsians." *Pharmacy in History* 29, no. 1 (1987): 3–28.

——— "Tashrih." *Encyclopaedia of Islam*. 2nd edition, 10: 354–356. Leiden: Brill, 2000.

——— "Anatomical Illustrations in Arabic Manuscripts." In *Arab Painting: Text and Image in Illustrated Arabic Manuscripts*, edited by Anna Contadini, 147–159. Leiden; Boston: Brill, 2007.

Scalenghe, Sara. "The Deaf in Ottoman Syria, 16th–18th Centuries." *Arab Studies Journal* 13 (2005): 10–25.

Schacht, Joseph. *An Introduction to Islamic Law*. Oxford: Clarendon Press, 1964.

Schimmel, Annemarie. *Islamic Names*. Edinburgh: Edinburgh University Press, 1989.

Schmidt, Jeremy. *Melancholy and the Care of the Soul: Religion, Moral Philosophy, and Madness in Early Modern England*. Aldershot; Burlington, VT: Ashgate, 2007.

Schweik, Susan M. *The Ugly Laws: Disability in Public*. New York: New York University Press, 2009.

Sedra, Paul. *From Mission to Modernity: Evangelicals, Reformers and Education in Nineteenth-Century Egypt*. New York: I. B. Tauris, 2011.

Seikaly, Samir M. "Damascene Intellectual Life in the Opening Years of the 20th Century: Muhammad Kurd 'Ali and *al-Muqtabas*." In *Intellectual Life in the Arab East, 1890–1939*, edited by Marwan R. Buheiry, 125–153. Beirut: Center for Arab and Middle East Studies, American University of Beirut, 1981.

Shahin, Hashem et al. "Genetics of Congenital Deafness in the Palestinian Population: Multiple Connexin 26 Alleles with Shared Origins in the Middle East." *Human Genetics* 110 (2002): 284–289.

Shakespeare, Tom. *Disability Rights and Wrongs*. New York: Routledge, 2006.

Sharpe, Robert L. "England's Legal Monsters." *Law, Culture, and the Humanities* 5, no. 1 (2009): 100–130.

Shefer-Mossensohn, Miri. "Physicians in Mamluk and Ottoman Courts." *In Mamluk and Ottomans: Studies in Honour of Michael Winter*, edited by David J. Wasserstein and Ami Ayalon, 114–122. London; New York: Routledge, 2006.

——— *Ottoman Medicine: Healing and Medical Institutions, 1500–1700*. Albany: State University of New York Press, 2009.

Siebers, Tobin. *Disability Theory*. Ann Arbor: The University of Michigan Press, 2008.

Sirriyeh, Elizabeth. *Sufi Visionary of Ottoman Damascus: 'Abd al-Ghani al-Nabulusi, 1641–1731*. New York: RoutledgeCurzon, 2004.

Skovgaard-Petersen, Jakob. "Sex Change in Cairo: Gender and Islamic Law," *The Journal of the International Institute* 2, no. 3 (Spring 1995). Available online at http://quod.lib.umich.edu/j/jii/4750978.0002.302/–sex-change-in-cairo-gender-and-islamic-law?rgn=main;view=fulltext. Accessed October 8, 2013.

Smith, Bonnie G., and Beth Hutchinson, eds. *Gendering Disability*. New Brunswick, NJ: Rutgers University Press, 2004.

Snyder, Sharon L., Brenda Jo Brueggemann, and Rosemarie Garland-Thomson, eds. *Disability Studies: Enabling the Humanities*. New York: Modern Language Association of America, 2002.

Snyder, Sharon L., and David T. Mitchell. *Cultural Locations of Disability*. Chicago: Chicago University Press, 2006.

Stainton, Tim. "Reason, Grace, and Charity: Augustine and the Impact of Church Doctrine on the Construction of Intellectual Disability." *Disability and Society* 23, no. 5 (2008): 485–496.

Stephan, Stephan H. "Lunacy in Palestinian Folklore." *Journal of the Palestine Oriental Society* 5 (1925): 1–16.

Stiker, Henri-Jaques. *A History of Disability*. Translated from French by William Sayers. Ann Arbor: The University of Michigan Press, 1999.

Stone, Deborah A. *The Disabled State*. Philadelphia: Temple University Press, 1984.

Sufian, Sandra M. *Healing the Land and the Nation: Malaria and the Zionist Project in Palestine, 1920–1947*. Chicago: University of Chicago Press, 2007.

Al-Tabba', 'Uthman Mustafa. *Ithaf al-A'izza fi Tarikh Ghazza*. Edited by 'Abd al-Latif Zaki Abu Hashim. 4 vols. Gaza: Maktabat al-Yaziji, 1999.

Tadmouri, Ghazi O. et al. "Consanguinity and Reproductive Health among Arabs." *Reproductive Health* 6:17 (2009): n.p.

Tamari, Salim. "Lepers, Lunatics and Saints: The Nativist Ethnography of Tawfiq Canaan and His Jerusalem Circle." *Jerusalem Quarterly* 20 (2004): 24–43.

Tamari, Stephen E. "Biography, Autobiography, and Identity in Early Modern Damascus." In *Auto/Biography and the Construction of Identity and Community in the Middle East*, edited by Mary Ann Fay, 37–49. New York: Palgrave, 2001.

Tayel, Shawky M., et al. "Congenital Adrenal Hyperplasia in Alexandria, Egypt: A High Prevalence Justifying the Need for a Community-Based Newborn Screening Program." *Journal of Tropical Pediatrics* 57, no. 3 (2011): 232–234.

Teebi, Ahmad S., ed. *Genetic Disorders among Arab Populations*. 2nd edition. Berlin: Springer, 2010.

Temtamy, S. A., Meguid N. Abdel, I. Mazen, S. Ismail, S. Kassem, and R. Bassiouni. "A Genetic Epidemiological Study of Malformations at Birth in Egypt." *Eastern Mediterranean Health Journal* 4 (1998): 252–259.

Temtamy, Samia A., Mona S. Aglan, and Nagwa A. Meguid. "Genetic Disorders in Egypt." In *Genetic Disorders Among Arab Populations*, 2nd edition, edited by Ahmad S. Teebi, 245–247. Berlin: Springer, 2010.

Thomson, Rosemarie Garland. *Extraordinary Bodies: Figuring Physical Disability in American Culture and Literature*. New York: Columbia University Press, 1997.

Tremain, Shelley, ed. *Foucault and the Government of Disability*. Ann Arbor: The University of Michigan Press, 2005.

Tucker, Judith E. *In the House of the Law: Gender and Islamic Law in Ottoman Syria and Palestine*. Berkeley: University of California Press, 1998.

_____ "'And God Knows Best': The Fatwa as a Source for the History of Gender in the Arab World." In *Beyond the Exotic: Women's Histories in Islamic Societies*, edited by Amira El-Azhari Sonbol, 165–179. Syracuse, NY: Syracuse University Press, 2005.

Turmusani, Majid. "Disabled Women in Islam: Middle Eastern Perspective." *Journal of Religion, Disability and Health* 5, no. 2/3 (2001): 73–85.

_____ *Disabled People and Economic Needs in the Developing World: A Political Perspective from Jordan*. Aldershot, UK: Ashgate, 2003.

Turner, David M. *Disability in Eighteenth-Century England: Imagining Physical Impairment*. New York; London: Routledge, 2012.

Turner, David M., and Kevin Stagg, eds. *Social Histories of Disability and Deformity*. London; New York: Routledge, 2006.

Turner, Wendy, ed. *Madness in Medieval Law and Custom*. Leiden: Brill, 2010.

Turner, Wendy, and Tory Vandeventer Pearman, eds. *The Treatment of Disabled Persons in Medieval Europe: Examining Disability in the Historical, Legal, Literary, Medical, and Religious Discourses of the Middle Ages*. Lewiston, NY: The Edwin Mellen Press, 2010.

Al-'Ulabi, Akram Hasan. *Takmilat Shadharat al-Dhahab fi Akhbar Man Dhahab: Tarajim al-A'lam fi al-Qarn al-Hadi 'Ashar al-Hijri fi al-'Alam al-Islami*. Damascus: Dar al-Tabba', 1991.

Ullmann, Manfred. *Islamic Medicine*. Translated from German by Jean Watt. Edinburgh: Edinburgh University Press, 1978.

van Gelder, Geert Jan H. "Hearing and Deafness." In *Encyclopaedia of the Qur'an*, edited by Jane McAuliffe, 2: 405–406. Leiden: Brill, 2002.

Van Cleve, John V., ed. *Deaf History Unveiled: Interpretations from the New Scholarship*. Washington, DC: Gallaudet University Press, 1993.

Walker, Nigel. *Crime and Insanity in England*. 2 vols. Edinburgh: Edinburgh University Press, 1968–1973.

Watenpaugh, Heghnar Zeitlian. "Deviant Dervishes: Space, Gender, and the Construction of Antinomian Piety in Ottoman Aleppo." *International Journal of Middle East Studies* 37, no. 4 (2005): 535–565.

Wehbi, Samantha. "The Challenges of Inclusive Education in Lebanon." *Disability and Society* 21, no. 4 (2006): 331–343.

Wehbi, Samantha, and Y. El-Lahib. "The Employment Situation of People with Disabilities in Lebanon: Challenges and Opportunities." *Disability and Society* 22, no. 4 (June 2007): 371–382.

Wehbi, Samantha, and Sylvana Lakkis. "Women with Disabilities in Lebanon: From Marginalization to Resistance." *Affilia* 25, no. 1 (2010): 56–67.

Weygand, Zina. *The Blind in French Society from the Middle Ages to the Century of Louis Braille*. Translated from French by Emily-Jane Cohen. Stanford, CA: Stanford University Press, 2009.

Wheatley, Edward. *Stumbling Blocks Before the Blind: Medieval Constructions of a Disability*. Ann Arbor: The University of Michigan Press, 2010.

Winter, Michael. *Society and Religion in Early Ottoman Egypt: Studies in the Writings of ʿAbd al-Wahhāb al-Shaʿrānī*. New Brunswick, NJ: Transaction Books, 1982.

______ *Egyptian Society under Ottoman Rule, 1517–1798*. London; New York: Routledge, 1992.

______ "Islamic Attitudes toward the Human Body." In *Religious Reflections on the Human Body*, edited by Jane Marie Law, 36–45. Bloomington: Indiana University Press, 1995.

______ "Historiography in Arabic during the Ottoman Period." In *Arabic Literature in the Post-Classical Period*, edited by Roger Allen and D. S. Richards, 171–188. Cambridge; New York: Cambridge University Press, 2006.

Ze'evi, Dror. *Producing Desire: Changing Sexual Discourses in the Ottoman Middle East, 1500–1900*. Berkeley: University of California Press, 2006.

Zeshan, Ulrike. "Sign Language in Turkey: The Story of a Hidden Language." *Turkic Languages* 6, no. 2 (2002): 229–274.

______ "Aspects of Türk İşaret Dili (Turkish Sign Language)." *Sign Language and Linguistics* 6, no. 1 (2003): 43–75.

______ "Sign Languages of the World." In *Encyclopedia of Language and Linguistics*, 2nd edition, edited by Keith Brown et al., 11: 358–365. Boston: Elsevier, 2006.

Ziadat, Adel A. *Western Science in the Arab World: The Impact of Darwinism, 1860–1930*. New York: St. Martin's Press, 1986.

Index

Titles in the Series

Popular Culture in Medieval Cairo Boaz Shoshan
Early Philosophical Shiism: The Ismaili Neoplatonism of Abū Ya'qūb Al-Sijistānī Paul E. Walker
Indian Merchants and Eurasian Trade, 1600–1750 Stephen Frederic Dale
Palestinian Peasants and Ottoman Officials: Rural Administration Around Sixteenth-Century Jerusalem Amy Singer
Arabic Historical Thought in the Classical Period Tarif Khalidi
Mongols and Mamluks: The Mamluk–Īlkhānid War, 1260–1281 Reuven Amitai-Preiss
Hierarchy and Egalitarianism in Islamic Thought Louise Marlow
The Politics of Households in Ottoman Egypt: The Rise of the Qazdağlis Jane Hathaway
Commodity and Exchange in the Mongol Empire: A Cultural History of Islamic Textiles Thomas T. Allsen
State and Provincial Society in the Ottoman Empire: Mosul, 1540–1834 Dina Rizk Khoury
The Mamluks in Egyptian Politics and Society Thomas Philipp and Ulrich Haarmann (eds.)
The Delhi Sultanate: A Political and Military History Peter Jackson
European and Islamic Trade in the Early Ottoman State: The Merchants of Genoa and Turkey Kate Fleet
Reinterpreting Islamic Historiography: Hārūn Al-Rashīd and the Narrative of the 'Abbāsid Caliphate Tayeb El-Hibri
The Ottoman City Between East and West: Aleppo, Izmir, and Istanbul Edhem Eldem, Daniel Goffman, and Bruce Masters
A Monetary History of the Ottoman Empire Şevket Pamuk
The Politics of Trade in Safavid Iran: Silk for Silver, 1600–1730 Rudolph P. Matthee
The Idea of Idolatry and the Emergence of Islam: From Polemic to History G. R. Hawting
Classical Arabic Biography: The Heirs of the Prophets in the Age of Al-Ma'mūn Michael Cooperson
Empire and Elites After the Muslim Conquest: The Transformation of Northern Mesopotamia Chase F. Robinson
Poverty and Charity in Medieval Islam: Mamluk Egypt, 1250–1517 Adam Sabra
Christians and Jews in the Ottoman Arab World: The Roots of Sectarianism Bruce Masters
Culture and Conquest in Mongol Eurasia Thomas T. Allsen
Revival and Reform in Islam: The Legacy of Muhammad Al-Shawkani Bernard Haykel
Tolerance and Coercion in Islam: Interfaith Relations in the Muslim Tradition Yohanan Friedmann
Guns for the Sultan: Military Power and the Weapons Industry in the Ottoman Empire Gábor Ágoston
Marriage, Money and Divorce in Medieval Islamic Society Yossef Rapoport

The Empire of the Qara Khitai in Eurasian History: Between China and the Islamic World Michal Biran
Domesticity and Power in the Early Mughal World Ruby Lal
Power, Politics and Religion in Timurid Iran Beatrice Forbes Manz
Postal Systems in the Pre-Modern Islamic World Adam J. Silverstein
Kingship and Ideology in the Islamic and Mongol Worlds Anne F. Broadbridge
Justice, Punishment and the Medieval Muslim Imagination Christian Lange
The Shiites of Lebanon Under Ottoman Rule, 1516–1788 Stefan Winter
Women and Slavery in the Late Ottoman Empire Madeline Zilfi
The Second Ottoman Empire: Political and Social Transformation in the Early Modern World Baki Tezcan
Non-Muslims in the Early Islamic Empire: From Surrender to Coexistence Milka Levy-Rubin
The Legendary Biographies of Tamerlane: Islam and Heroic Apocrypha in Central Asia Ron Sela
Law and Piety in Medieval Islam Megan Reid
The Origins of the Shiʿa: Identity, Ritual, and Sacred Space in Eighth-Century Kufa Najam Haider
Politics, Law, and Community in Islamic Thought: The Taymiyyan Moment Ovamir Anjum
The Power of Oratory in the Medieval Muslim World Linda G. Jones
Animals in the Qurʾan Sarra Tlili
The Logic of Law Making in Islam: Women and Prayer in the Legal Tradition Behnam Sadeghi
Empire and Power in the Reign of Süleyman: Narrating the Sixteenth-Century Ottoman World Kaya Şahin
Women and the Transmission of Religious Knowledge in Islam Asma Sayeed
The New Muslims of Post-Conquest Iran: Tradition, Memory, and Conversion Sarah Bowen Savant
The Mamluk City in the Middle East: History, Culture, and the Urban Landscape Nimrod Luz